**Fossil Vertebrates
of Africa**

Volume 1

Fossil Vertebrates of Africa

Edited by

L. S. B. Leakey

Volume 1

1969 ACADEMIC PRESS

NEW YORK AND LONDON

ACADEMIC PRESS INC. (LONDON) LTD

Berkeley Square House,
Berkeley Square,
London, W1X 6BA

U.S. Edition published by

ACADEMIC PRESS INC.

111 Fifth Avenue,
New York, New York 10003

Library of Congress Catalog Card Number: 71-92394
SBN: 12-440401-4

PRINTED IN GREAT BRITAIN BY
THE WHITEFRIARS PRESS LTD., LONDON AND TONBRIDGE

CONTRIBUTORS

P. M. BUTLER   Professor of Zoology in Royal Holloway College, London, England.

D. A. HOOIJER   Rijksmuseum van Natuurlijke Historie, Leiden, Holland.

G. H. R. VON KOENIGSWALD   Senckenberg Museum, Frankfurt am Main, Germany.

R. E. F. LEAKEY   Centre for Prehistory and Palaeontology, Nairobi, Kenya.

PREFACE

In 1950 arrangements were made with the Director of the British Museum of Natural History, in London, to publish a series of reports under the general title of Fossil Mammals of Africa. The first number in the series appeared in 1951, entitled " Miocene Hominoidea of East Africa " by W. Le Gros Clark and L. S. B. Leakey. Since then a further 21 numbers have been issued (see List of Titles and Authors inside last page of this volume). The Trustees of the British Museum have now decided to cease publishing this series. This is because in most cases these reports do not now include studies of any material which is (or which will become in due course), their property.

In earlier days the Governments of British Overseas Colonies and Territories accepted that all " Type " specimens of new fossil species, as well as a proportion of all other material, should be handed over to what was then the main Commonwealth Museum of Natural History. Nowadays the Governments of former colonies quite rightly and naturally insist that all fossil material excavated in their countries is National property and that, after study and publication, it must be sent to the relevant National Museum. In consequence the British Museum of Natural History no longer benefits from such scientific excavations.

Arrangements have now been made between the Director of the Centre for Prehistory and Palaeontology, Nairobi, and the Academic Press London to resume production of the series under the title " Fossil Vertebrates of Africa ".

This volume is the first to be published under the new arrangement.

As far as possible, the format will follow closely that of the volumes published before, but they will have hard covers and, in some cases, several different papers will appear in a single number, as in the case of this first volume.

The reports will be by a wide variety of authors in different countries, since material is being submitted for study and report to those most competent to carry out the work at any given time.

Since palaeontological research is growing rapidly in Africa, it may be expected that this series will continue for a very long time. The majority of the reports will deal with genera and species of fossil mammals new to Science, but in some cases fossil reptiles and birds will be included.

Only the absolute minimum of editing will be undertaken, so as to leave each report (as far as possible) exactly as written by its author.

April 1969

L. S. B. LEAKEY,
Editor.

CONTENTS

INSECTIVORES AND BATS FROM THE MIOCENE OF EAST AFRICA: NEW MATERIAL

P. M. BUTLER

Royal Holloway College, University of London, Englefield Green, Surrey, England

SYNOPSIS

Sixty three specimens of insectivores and bats, previously unreported, are recorded from Rusinga, Songhor, Koru and Napak. *Rhynchocyon rusingae*, new species, is described. Additional information is provided on the dentition of *R. clarki*, and two specimens show the existence of unnamed species of Macroscelididae. The genus *Lanthanotherium* is recognized in the fauna. More material of *Gymnurechinus songhorensis* adds to knowledge of

1

its dentition. New information on *Protenrec tricuspis* relates to the upper cheek teeth, the anterior margin of the orbit and the lower premolars and canine. A cranium and lower teeth referred to *Erythrozootes chamerpes* are described. A new subfamily, Protenrecinae, is proposed for *Protenrec* and *Erythrozootes*, and the relationships of the Tenrecidae are discussed with special reference to the Miocene forms. Further information is provided on *Prochrysochloris miocaenicus*, including the lower dentition. A bat humerus is placed in the Emballonuridae, and another constitutes the oldest record of the Hipposideridae from Africa.

INTRODUCTION

The present paper is written as a supplement to previous reports (Butler, 1956b, Butler and Hopwood, 1957). A number of additional specimens have since come into my hands, mainly from Rusinga and Songhor, sent by Dr. L. S. B. Leakey, and some from Napak, collected by Dr. W. W. Bishop. Some specimens from Koru, now in the British Museum, are also included in the present report. Though most of the new specimens are referable to species already known, many of them provide additional information on morphology and distribution.

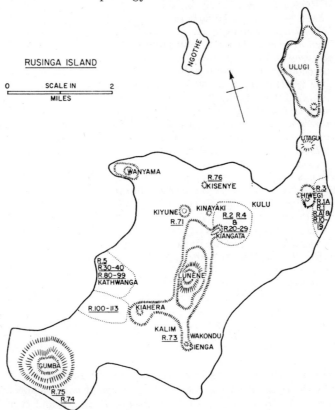

FIG. 1. Map of Rusinga Island showing the fossiliferous localities. From Le Gros Clark and Leakey (1951).

Most of the specimens are the property of the Centre for Prehistory and Palaeontology, Nairobi, and are referred to by their field collection numbers: thus Rs 891.56 is no. 891 of the Rusinga collection of 1956. The remaining specimens are the property of the British Museum (Natural History).

The material comes from four localities: Rusinga, Songhor and Koru, in the Kavirondo Gulf area of Kenya, and Napak, about 200 miles to the north in Uganda. The geology has been reviewed by Bishop (1965), who emphasized the similarity between the faunal lists of these localities, indicating that they are roughly of similar age. This has been generally considered to be broadly equivalent to the Burdigalian of Europe, but it is by no means certain.* At Rusinga, where the deposits reach a considerable thickness, the stratigraphy has not been completely elucidated (see Shackleton, 1951; Kent, 1945), and it is possible that part of the fauna is of later date. A map showing the distribution of the fossiliferous sites on Rusinga is given by Whitworth (1954) and is reproduced here (fig. 1). For an account of the Napak sites see Bishop (1964).

As will be seen from Table I, the majority of insectivores from Rusinga come from the Hiwegi series (sites R1, R1a, R3, R3a, R12), but 24 specimens from the Kathwanga area may be older (Kiahera series) or younger (Kathwanga series). If species known only by single specimens are ignored, six species have been found both at Rusinga and Songhor: *Rhynchocyon clarki*, *R. rusingae*, *Galerix africanus*, *Gymnurechinus songhorensis*, *Amphechinus rusingensis*, *Protenrec tricuspis*. These six species comprise 75% of the specimens obtained from Songhor but only 40% of those from Rusinga. The difference is probably one of facies: *Gymnurechinus leakeyi*, the commonest species at Rusinga, is absent from Songhor, where it is replaced by *G. songhorensis*, a rare species at Rusinga; *Protenrec tricuspis* is represented by six specimens from Songhor but only one from Rusinga; *Erythrozootes chamerpes* and *Prochrysochloris miocaenicus* are unknown from Rusinga. The insectivore faunas of Koru and Napak are too poorly known for meaningful comparison: out of four species from Koru, two occur at Songhor; the two species from Napak both occur at Songhor, one also at Koru and the other also at Rusinga.

SYSTEMATIC DESCRIPTIONS

Order MACROSCELIDEA Patterson, 1965

In 1956a the author made the suggestion that the elephant shrews should be placed in a separate order. Patterson (1965) first formally proposed the order and gave a definition.

Family *MACROSCELIDIDAE* Mivart, 1868

The fossil members of this family, known only from Africa, have been reviewed by Patterson (1965). The only species hitherto described from the Miocene are

*Since this was written Van Couvering and Miller (1969) have published an account of the stratigraphy of Rusinga which clears up the earlier uncertainties.

TABLE I

Tally of independent specimens of Insectivores and Bats from "Burdigalian" sites in East Africa

(excluding *Myohyrax*)

	Napak I, IV	Koru	Songhor	Rusinga				
				L. Hiwegi series R1, R1a, R12, Wanyama	U. Hiwegi series R3, R3a	Kulu series R2	L. Kathwanga or Kiahera series R105a, R106, R111	Site unknown
Rhynchocyon clarki	—	—	5	4	4	1	2	—
R. rusingae	—	—	4	1	5	—	1	—
Macroscelididae sp. 1	—	—	—	1	—	—	—	—
Macroscelididae sp. 2	—	1	—	—	—	—	—	—
Galerix africanus	—	—	1	—	—	—	—	—
Lanthanotherium sp.	—	—	4	—	1	—	—	—
Gymnurechinus leakeyi	—	—	—	10	17	6	10	2
G. camptolophus	—	—	—	—	1	—	8	2
G. songhorensis	—	—	10	—	—	—	—	—
Amphechinus rusingensis	—	—	4	2	1	1	3	—
Crocidura sp.	—	—	—	1	1	1	—	—
Protenrec tricuspis	2	—	6	—	—	—	—	—
Erythrozootes chamerpes	1	2	2	—	—	—	—	—
Geogale aletris	—	—	3	—	1	—	—	—
Prochrysochloris miocaenicus	—	1	—	—	—	—	—	—
Saccolaimus incognita	—	1	—	—	—	—	—	—
Emballonuridae sp.	—	—	—	—	1	—	—	—
Megadermidae sp.	—	—	—	—	—	—	1	—
?Hipposideros sp.	—	—	1	—	—	—	—	—
Total species	2	4	10	6 or 7	7 or 8	3	6	2
Total specimens	3	5	40	19 or 20	30 or 31	8	25	4

Total Rusinga: species 13
specimens 87

Rhynchocyon clarki Butler and Hopwood and two species of Myohyracinae, transferred from the Hyracoidea to the Macroscelidea by Patterson.

Genus *Rhynchocyon* Peters, 1847

The presence of this genus in the Miocene was recognized on the basis of the anterior part of a skull and five mandibular fragments. Twenty additional specimens of teeth and jaws from Rusinga and Songhor show from their diversity of size and structure that more than one species is represented.

Cf. *Rhynchocyon clarki* Butler and Hopwood, 1957

Additional material.—Rusinga: Rs 891.56, right maxillary fragment with M^2 and M^1, from site R3; Rs 719.47, right maxillary fragment with M^1 and P^4, from site R3; Rs 627.56, two associated incomplete mandibular rami, with M_2–P_4 and unerupted P_3 on both sides and unerupted P_2 on the left side, from site R2; Rs 1970.50, fragment with left P_4 from Kathwanga; Rs 587.49, fragment with right Pd_4 from site R3a.

Songhor: Sgr 3160f.66, fragment with left M_1 and P_4; Sgr 3119.66, fragment with unerupted left M_2; Sgr 1814.66, fragment with unerupted left P_4; Sgr 3160d.66, fragment with poorly preserved left M_1.

Reference of this material to *R. clarki* is based mainly on size, as the teeth cannot be directly compared with the holotype. Measured from the alveoli, the total length of M^2–P^2 is 14·5 mm in the holotype, which compares with a total length of M_2–P_2 of 14·9 mm in Rs 627.56. In Recent species of *Rhynchocyon* M^2–P^2 is nearly equal in length to M_2–P_2.

Description.—The new specimens provide the following additional information:—
M^2 is well preserved in Rs 891.56 (fig. 2a). Its posterior part is much reduced as compared with M^1, though it bears a well-developed metacone, nearly as high as the paracone but much shorter; the hypocone is represented only by a short ridge which branches off the protocone-metacone ridge and runs posteriorly, forming the border of a small basin on the lingual side of the metacone.

The anterior ridge of the entoconid of M_2 in Sgr 3119.66 is developed into a small basal cusp, or entostylid, seen also in Rs 1700.50. The development of this cusp seems to be variable as it is absent from Rs 627.56.

P_4 (fig. 2c) differs from M_1 mainly in the greater length of the trigonid, the paraconid being more widely separated from the other trigonid cusps. An anterobuccal cingulum is present below the paraconid, and a vertical groove passes up the anterobuccal surface between the protoconid and the paraconid. The best specimens are Sgr 1814.66 and Rs 627.56. The length of the tooth is rather variable (range 3·6–4·3 mm), mainly due to the degree of anterior prominence of the paraconid.

P_3 is quite unworn in Rs 627.56. It is much shorter than P_4, and its protoconid would probably be slightly lower when the tooth was fully erupted. The paraconid is a poorly differentiated anterior cusp, only one-third of the height of the protoconid, measured from cingulum level. There is a very short length of anterobuccal cingulum and another short piece of lingual cingulum near the paraconid. A sharp ridge runs from the tip of the protoconid to the posterior end of the tooth; its upper part has a slightly lingual deviation, but there is no metaconid. A small posterior basal cusp, together with a short length of lingual cingulum, represents the talonid. P_2 of this specimen is embedded in the jaw but appears to be similar to P_3 though smaller in size. Nothing is known of the dentition anterior to P_2.

Rs 587.49 shows a worn Pd_4 (fig. 2d) associated with the alveoli of M_1 which from their size appear to belong to this species. Beneath is an unerupted P_4. Pd_4 is narrower and lower than P_4 but somewhat longer. The talonid is molariform, except that it is proportionately narrower. There is a small entostylid on the anterior slope of the entoconid. The trigonid is extended anteriorly even more than in P_4; the paraconid tip is situated towards the lingual side of the tooth, and it is connected to the protoconid by a long crest, below which is an anterobuccal cingulum.

Rhynchocyon rusingae sp. nov.

Diagnosis.—Resembling *R. clarki* but larger in size (M_2–P_2 of holotype 21·2 mm), and possessing a protostylid on P_3 and P_2.

Holotype.—Rs 711.47, an incomplete left mandible with P_4 and P_3, and an associated right mandibular fragment without teeth, from Rusinga, site R3.

Paratypes.—Sgr 362.49, a right mandibular fragment with M_2 and M_1; Sgr 81.62, a left mandibular fragment with Pd_4 and Pd_3; Sgr 3120b.66, an isolated M^1; all from Songhor.

Referred specimens.—Rusinga: Rs 68.51, fragment with left M_1, from site 3a; Rs 439.51, fragment with parts of left M_1 and P_4, from site 3a; Rs 778.47, fragment with left P_4, from site R106; Rs 788.56, fragment with badly preserved P_3 and P_2, from site R1; Rs 579.56, fragment with right P_3 and P_2, from site R3; Rs 582.56, fragment with right Pd_4 and Pd_3, from site R3.

Songhor: Sgr 3160b, an isolated Pd_4.

Description.—The only known upper tooth is M^1, represented by a single but almost unworn specimen (Sgr 3120b.66) (fig. 2b). It resembles specimens of M^1 referred to *R. clarki* except in its larger size (measurements: 3·7 × 3·8 mm): it is much too large to fit the roots of M^1 of the holotype of *R. clarki*.

The holotype of *R. rusingae* contains P_4 and P_3, and also all the alveoli as far forward as P_1 (fig. 3e). The total length of P_2–M_2 is 21·2 mm, 46% longer than P^2–M^2 in the holotype of *R. clarki*.

Sgr 362.49 contains M_2 and M_1 (fig. 3a), agreeing in size with the alveoli of the holotype and much larger than lower molars referred to *R. clarki*. The teeth are worn, and the entoconid of M_2 has been broken off. Rs 439.51 shows the alveoli of M_2, a worn and incomplete M_1 and part of the talonid of P_4. Rs 68.51 has a complete but worn M_1. No differences of pattern from *R. clarki* can be detected in these worn molars.

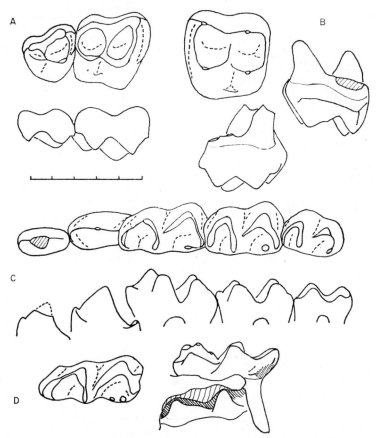

Fig. 2. *Rhynchocyon*. A, Cf. *R. clarki*, Rs 891.56, M² and M¹, crown and buccal views. B, *R. rusingae*, Sgr 3120b.66, M¹, crown, anterior and buccal views. c, Cf. *R. clarki*, Rs 627.56, P₂–M₂, crown and lingual views. D, Cf. *R. clarki*, Rs 587.49, Pd₄, crown view and buccal view (showing unerupted P₄ below). In this and other figures the scale represents 5 mm.

P₄ of the holotype (fig. 3a) has suffered some damage due to weathering, and the tips of the cusps appear to have been worn off, but its morphology is clear. It is considerably larger than P₄ of *R. clarki* as seen in Rs 627.56, and it also differs from that specimen in that the paraconid projects more anteriorly and is marked off from the protoconid by a deeper groove on the buccal side. In Rs 778.47 the elongation of the trigonid is not so marked.

P₃ of the holotype is well preserved except for some weathering (fig. 3a, 3b). The paraconid is a much more distinct cusp than in *R. clarki*, about half the height of the protoconid. A ridge passes from the tip of the paraconid over the tip of the protoconid and down its posterior slope. When seen in crown view this ridge is straight. Posteriorly to the protoconid it rises to form a small but distinct accessory

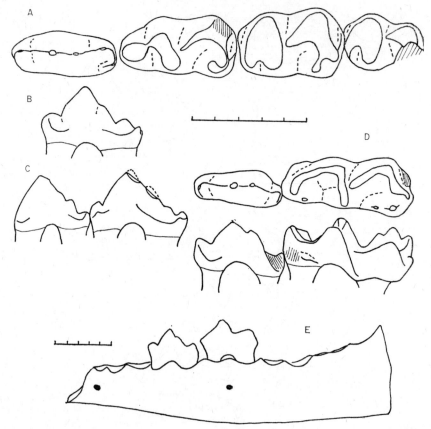

FIG. 3. *Rhynchocyon rusingae*. A, P_3 and P_4 of Rs 711.47 (reversed) and M_1 and M_2 of Sgr 362.49. B, Rs 711.47, P_3, lingual view (reversed). C, Rs 579.56, P_2 and P_3, lingual view. D, Sgr 81.62, Pd_3 and Pd_4, crown and reversed buccal views. E, Rs 711.47, outer view of mandible.

cusp (protostylid), and near the posterior end of the tooth it connects with the hypoconid. This is a small cusp, lower than the paraconid and displaced slightly towards the buccal side; on its lingual side there is a rudimentary talonid basin, and a small entoconid is present at the posterolingual edge of the crown. Thus the talonid is better developed than in *R. clarki*. Rs 579.56 (fig. 3c) shows some differences: the paraconid is smaller; small cuspules are developed on the cingulum buccally and lingually near the paraconid; the posterior ridge of the protoconid bears two cusps, together with a third elevation at its posterior end (either a hypoconid or a hypoconulid). Whether these differences are of taxonomic value is unknown.

P_2 is represented in the holotype only by its alveoli. In Rs 579.56 it is similar in structure to P_3 but smaller and somewhat simplified: the paraconid is absent and the protostylid is less distinct, though present. Lower down on the posterior ridge is a second cusp like that on P_3 of this specimen, followed by a posterior basal cusp.

The entoconid, present on P_3, is not developed on P_2, and the talonid basin is very rudimentary. In the holotype there are two alveoli of equal size anterior to P_2, probably for a two-rooted P_1.

Two mental foramina are visible in the holotype, below P_4 and P_1. The symphysis ends below P_1.

Three specimens of Pd_4 are referred to this species because of their relatively large size (fig. 3d). They agree in structure with Rs 587.49, referred to *R. clarki*, but in two of them (Sgr 81.62 and Sgr 3160b.66) an entostylid is present on the anterior slope of the entoconid; in the third specimen the entoconid is broken.

Pd_3 is preserved in Rs 582.56 and Sgr 81.62 (fig. 3d). It is much lower than P_3, and it has a more strongly-developed protostylid, displaced somewhat buccally. The paraconid is well developed. At the posterior end there is a small basal cusp, but the talonid basin and the entoconid are rudimentary. Both specimens show a strong groove of wear on the buccal side of the talonid, due to the paracone of Pd^3. From the alveoli in Rs 582.56, Pd_2 was a two-rooted tooth somewhat shorter than Pd_3.

Discussion.—*R. rusingae* approaches the Recent species of *Rhynchocyon* in the possession of a protostylid on P_3 and P_2. In Recent species the size of the protostylid varies, and it may be very rudimentary on P_3 and absent on P_2, especially in *R. chrysopygus* and *R. petersi*; it is always present on Pd_3. The protostylid does not occur in Recent members of the Macroscelidinae, but in these there is a posterolingual crest on Pd_3 which bears a metastylid near its lower end. The metastylid is so identified because (1) it bites between the protoconule and the protocone, like the metastylid of Pd_4, and (2) it is sometimes connected by a transverse ridge to the hypoconid, again as in Pd_4. The protostylid of *Rhynchocyon* shears against the lingual side of the paracone-parastyle crest, its tip passing lingually to the notch between these two cusps.

R. rusingae is significantly larger than *R. clarki*, and approaches the Recent species of *Rhynchocyon* in size. The two species occur together at Songhor and Rusinga (sites R1, R3 and R3a).

Pd_4 in the Recent species differs from that of both Miocene species in the presence of a metastylid, closely posterior to the metaconid and connected to the hypoconid by an oblique crest. In the Miocene forms the hypoconid crest stops snort near the middle of the protoconid-metaconid crest. Similar differences occur on M_1.

The lower jaws with milk teeth referred to *Ptolemaia* by Schlosser (1911), but probably wrongly (Matthew, 1918), show a rather close resemblance to *Rhynchocyon rusingae*: the two-rooted Pd_1; the structure of Pd_3, which possesses a protostylid and a rudimentary talonid but no metaconid; the enlarged, forwardly placed paraconid of Pd_4; the lingual position of the comparatively large paraconids of M_1 and M_2, though unlike *Rhynchocyon* the paraconid is higher than the metaconid on these teeth. M_2 of " *Ptolemaia* " is however as well developed as M_1, and in B.M. M10189 an incompletely developed M_3 is present in a crypt. The presence of M_3 is not unexpected in an early member of the Macroscelididae: it is retained in *Myohyrax* and also in *Nasilio* (if its presence in this genus is not due to re-development). However, " *Ptolemaia* " differs from *Rhynchocyon* in its much larger canine, its deeper masseteric fossa and its more extensive symphysis which reaches to below the anterior end of Pd_3, as well as in the greater size and stouter construction of its mandible. The resemblances in the cheek teeth are probably due to convergence. The relationships of " *Ptolemaia* " have been recently discussed by Van Valen (1966), who suggested a relationship to the Pantolestidae.

Macroscelididae of Uncertain Genus and Species

M14282 (fig. 4a) consists of a right mandibular fragment with two teeth, interpreted as P_2 and P_3, from Koru (Leakey's Red Bed, in *situ*; for stratigraphy see Kent, 1945). The teeth are unworn. P_3 is 3·2 mm long, thus comparable in size with *Rhynchocyon clarki*. It is slightly broadened posteriorly. It differs from *Rhynchocyon* in the possession of a metaconid, posterolingual to the protoconid and connected to that cusp by a ridge. A low but sharp ridge runs from the buccal surface of the metaconid to the small talonid cusp. A more buccal ridge arises near the metaconid and runs posterobuccally to end in a small elevation near the margin of the tooth. There is no entoconid. The paraconid is well developed and anterior in position. P_2 is a smaller tooth, 2·8 mm long. It also possesses a metaconid, in the form of a minute cusp on the posterolingual slope of the protoconid. In addition, the posterior ridge of the protoconid is elevated to form a protostylid. The talonid cusp is rudimentary. The paraconid is again strongly developed.

These teeth are interpreted as premolars rather than milk molars because of the height of the crown. They differ from *Rhynchocyon* in the possession of a metaconid, and from *Myohyrax* in being much less hypsodont. The specimen evidently represents an unknown genus, but it is best left unnamed until more material is discovered.

Rs 514.49, from Rusinga site R1a, consists of two associated mandibular rami containing unworn milk molars (right Pd_4–Pd_2, left Pd_4 and Pd_3) (fig. 4B). It is comparable in size with *Rhynchocyon clarki*. Pd_4 has the usual macroscelidid pattern, with a molariform talonid and strongly-developed anterior paraconid. There is an additional talonid cusp (entostylid) anterior to the entoconid, as in Miocene specimens of *Rhynchocyon*. The paraconid is complex: there is an additional cusp on its posterior slope; the anterobuccal cingulum takes the form of a vertical ridge which rises to meet the paraconid-protoconid crest at a position immediately buccal to the paraconid, and a rather indistinct elevation of the crest is present at the point of junction. Accessory cusps posterior and buccal to the paraconid occur in some Recent specimens of *Rhynchocyon*. Pd_3, however, differs from *Rhynchocyon* in several respects. There is a trigonid in which the paraconid and metaconid are displaced

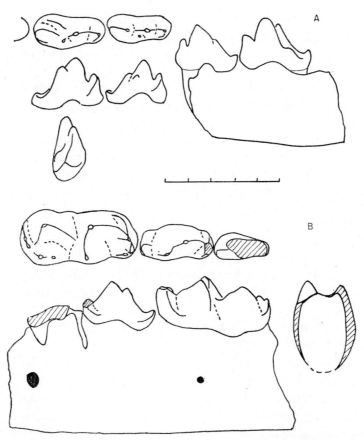

FIG. 4. Macroscelididae of uncertain genus and species. A, B.M. M14284, P_3 and P_2, crown, lingual and buccal views, and P_3 posterior view. B, Rs 514.49, Pd_4–Pd_2, crown and buccal views, and unerupted M_1 seen in broken section through the trigonid.

towards the lingual side. The paraconid shows some resemblance to that of Pd_4, a ridge passing up its buccal slope. The metaconid is posterolingual to the protoconid, the two cusps being connected by a ridge; no protostylid is present, unless the metaconid is considered as a protostylid displaced lingually. The small low talonid is sharply marked off from the trigonid, and it bears a medianly placed hypoconid and a more lingual entoconid, both very small cusps. Pd_2 is shorter than Pd_3; its trigonid is broken, but it is possible to see that it had a small, one-cusped talonid. Three alveoli are present anterior to Pd_2, but the anterior end of the jaw is incomplete and it is uncertain how many teeth were present. However, the roots of the anterior teeth appear to have been crowded and procumbent. The symphysis extends back to the posterior end of Pd_2, thus further back than in *Rhynchocyon*. At the posterior end of the fragment is part of an unerupted M_1, broken vertically through the back of the trigonid so that the tips of the protoconid and the metaconid are visible. The

height of the protoconid is 3·3 mm, much greater than in the little worn specimen of
R. clarki (Rs 627.56), which measures 2·2 mm. There is a mental foramen under
Pd_2 and a smaller one under Pd_4. Lengths of the teeth are as follows: Pd_4 5·0 mm,
Pd_3 3·3 mm, Pd_2 2·5 mm.

Pd_4 of this specimen resembles Pd_4 of *Myohyrax oswaldi* Andrews (Whitworth,
1954, fig. 13) in that the anterobuccal cingulum takes the form of a vertical ridge
passing up to the paraconid, and in the presence of an entostylid anterior to the
entoconid (the interpretation of Patterson (1965) is followed here). The height of the
crown of M_1 and the probably crowded and procumbent anterior teeth also suggest
Myohyrax, and the specimen is probably referable to the Myohyracinae. As so little
is known of the milk dentition of *Myohyrax*, further comparison is impossible. Pd_4
of the present specimen is slightly larger than that of *Myohyrax*, which measures
4·6 mm in length.

Order LIPOTYPHLA Haeckel, 1866

This name is preferred to Insectivora, which has come to have a wider connotation,
embracing Leptictidae, Pantolestidae and other families of primitive Eutheria.

Family *ERINACEIDAE* Bonaparte, 1838
Lanthanotherium sp.

Sgr 3116.66 (fig 5A) and Sgr 3114.66, from Songhor are fragments of right and left
mandibles, probably belonging to the same individual, which contain M_1, M_2 and
unerupted M_3. Sgr 1813.66, a left mandibular fragment with the molar alveoli and
part of the ascending ramus (fig. 5B), and Sgr 3117.66, a left fragment with alveoli
of M_3–P_4, may belong to the same species.

The total length of M_{1-3} is about 9 mm, as in *L. sansaniense* from Sansan (Filhol,
1891), and greater than in specimens from Viehhausen (7·8–8·2 mm; Seemann, 1938).
M_1 measures 3·6 × 2·5 mm. The trigonid is longer, higher and slightly narrower
than the talonid. The protoconid and metaconid though unworn are obtuse at the
tips; they are placed fairly closely together but are separated by a deep notch.
The metaconid is only slightly lower than the protoconid and a little more anterior
in position. The paraconid takes the form of a curved horizontal ridge which reaches
the lingual border of the tooth, and does not rise to a distinct point. The talonid
bears two cusps, the hypoconid and entoconid, of nearly equal height. They are
blunt at the tips, and their anterior sides are steep and rounded, without ridges. A
short horizontal ridge connects the bases of the hypoconid and protoconid, at a
height of 0·7 mm above the buccal cingulum, and in the valley between the metaconid
and the entoconid there is a minute cuspule. A lingual and slightly posterior ridge
on the hypoconid meets a buccal and slightly posterior ridge on the entoconid;
from the notch where these ridges meet, a short length of posterior cingulum falls
rapidly in passing behind the hypoconid. An anterobuccal cingulum continues

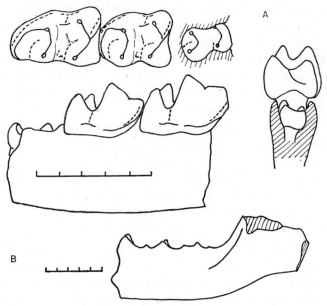

FIG. 5. *Lanthanotherium* sp. A, Sgr 3166.66, M_1–M_3, crown and buccal views, and M_2–M_3, posterior view. B, Sgr 1813.66, fragment of madible, outer view.

weakly round the buccal side of the protoconid and broadens to form a small shelf below the notch separating the protoconid and hypoconid. The base of the tooth markedly overhangs the roots on the buccal side.

M_2 measures 3.1×2.4 mm. The talonid and trigonid are of equal width, but the talonid forms only about two-fifths of the length of the tooth. The trigonid is shorter than in M_1 due to a somewhat weaker development of the paraconid, which, however, still reaches the lingual side of the tooth. In other respects M_2 resembles M_1.

M_3 is a much smaller tooth, incompletely exposed anteriorly. It resembles M_2 in pattern except that there is no cingulum posterior to the hypoconid.

Sgr 1813.66 shows molar alveoli of erinaceid type, including two alveoli for M_3. The alveolar length of M_3 measures 2.3 mm. The angle of elevation of the anterior border of the coronoid process is about 65° to the line of the teeth. Sgr 3117.66 also has two alveoli for M_3, but the alveolar length of this tooth is only 2.0 mm.

The only African species with which this material can be compared is *Galerix africanus* Butler. This species is based on a fragment containing P_3, P_4 and the alveoli of M_1 and M_2. Another fragment (Rs 1201.47), with M_2 and the alveoli of M_3, was referred to the same species. Although the tooth in Rs 1201.47 is worn, it is well enough preserved to permit a comparison with M_2 of the new material. The measurements agree, but M_2 of *Lanthanotherium* gives the impression of being relatively narrower, because its buccal and lingual cusps are placed more closely together. This may be the effect of a difference of wear, but there are other differences which cannot be ascribed to wear: in *Lanthanotherium* the cingulum round the protoconid is weaker than in *Galerix*, the talonid is shorter in proportion to the trigonid, the metaconid and entoconid are closer together, the entoconid lacks an anterior ridge present in *Galerix*, and the cingulum behind the hypoconid is shorter and steeper. It seems likely that these differences are of generic value. If Sgr 1813.66 is correctly referred to the same species as Sgr 3116.66, two further differences from *Galerix* may be noted: the coronoid process is less vertical (65° compared with 80°) and M_3 is slightly larger, judging from the alveoli.

The existence of two species of Echinosoricinae in the East African Miocene casts a doubt over the association of Rs 1201.47 with the holotype of *Galerix africanus*. This will be resolved only when molars and premolars are found together. For this reason I refrain from giving the new material a specific name.

The new material is referred to the genus *Lanthanotherium* because of the resemblance of the molars to *L. sansaniense*, notably in the short talonid, the wide trigonid angle and the character of the paraconid. Several species of the genus are known from Europe, mostly Vindobonian in age (see Viret, 1940; Seemann, 1938; Thenius, 1949) but extending into the L. Pliocene (Vallesian) in Spain (Villalta and Crusafont-Pairó, 1944). The genus has also been found in the U. Miocene and L. Pliocene of California (Webb, 1961; James, 1963).

Gymnurechinus leakeyi Butler, 1956

Thirteen additional specimens have been obtained from Rusinga:— M16353, a well-preserved cranium, from site R3; Rs 676.56, a crushed cranium, from site R2; Rs 1394.50, a well-preserved tip of the snout with roots of the teeth anterior to P^2, from Kathwanga; Rs 128.56, a fragment with the right upper canine, from site R2; Rs 53.56, right maxillary fragment with M^2–P^4, from site R2; Rs 1248.50, left maxillary fragment with M^1–P^3, from site R3a, s.c. Rs 1746.50, right maxillary fragment with P^4, P^3 and the alveoli of M^1, from site R3a; Rs 734.56, left maxillary fragment with M^2–P^4, from Kulu-Warengu; Rs 562.51, incomplete left mandible without teeth, showing the condyle, site unknown; Rs 789.56, fragment with unerupted left P_4, from site R1; Rs 312.51, anterior end of right mandible with P_4, from site R3; Rs 875.56, fragment of left mandible with M_1 and P_4, from site R3; Rs 504.51, fragment of right mandible with worn molar teeth, and an associated fragment with left M_3, from site R3. Except for the specimens from R2, all are from sites from which the species has already been reported. The known distribution of *G. leakeyi* is still restricted to Rusinga, from which 45 specimens have been obtained.

The only specimen of morphological interest is Rs 789.56, which shows a partly erupted P_4 next to the alveoli for M_1 (fig. 6a). P_4 resembles specimens referred to *G. camptolophus* in the independence of the paraconid, which is lower than the metaconid and separated from the protoconid by a deep notch. The tooth is, however,

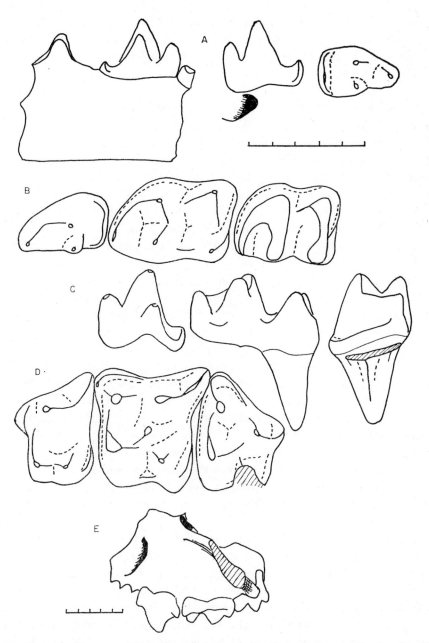

FIG. 6. *Gymnurechinus*. A, *G. leakeyi*, Rs 789.56, P_4, lingual view (with alveoli of M_1), buccal view (with mental foramen) and crown view. B–E, *G. songhorensis*. B, crown view of P_4 (Sgr 3974.66), M_1 (Sgr 2013a.66) and M_2 (Sgr 79.62, reversed). C, lingual view of P_4 (Sgr 3974.66) and lingual and anterior views of M_1 (Sgr 2013a.66). D, Sgr 408.66, P^4–M^2. E, Sgr 408.66, fragment of maxilla.

much smaller than in G. camptolophus (length 3·6 mm, compared with 3·9 mm in G. camptolophus), and it is interpreted as an aberrant specimen of G. leakeyi. It is unlikely to be a milk tooth as it erupts after M_1 and has a similar height to other specimens of P_4.

Gymnurechinus songhorensis Butler, 1956

Six additional specimens have been obtained from Songhor:—Sgr 408.66, a left maxillary fragment with M^2–P^4; Sgr 79.62, a left mandibular fragment with M_2 and M_1; Sgr 2013a.66, an isolated M_1; Sgr 3974.66, a fragment with left P_4; Sgr 3115. 66, fragment with the alveoli of M_{2-3}; Sgr 108.47, fragment with Pd_4.

In Sgr 408.66 P^4 and M^1 are well preserved, but on M^2 the hypocone has been broken away (figs. 6d, 6e). P^4 resembles the specimen previously described (Butler, 1956b, fig. 15) in size and proportions. M^1 differs from that specimen in that the metaconule is rather better developed. M^2 is a little larger than in G. leakeyi but similar in proportions and structure.

P_4 is well preserved in Sgr 3974.66 (figs. 6b, 6c). It measures 3·7 × 2·1 mm, only a little larger than in G. leakeyi (Rs 864.50). Its structure closely resembles G. camptolophus (Rs 595.48) in that the paraconid is a strongly-developed conical cusp, separated from the protoconid by a deep notch. The metaconid is lower than the paraconid, as in most other specimens of Gymnurechinus, but it is more widely separated from the protoconid than in Rs 864.50 and Rs 595.48. This tooth differs from Rs 789.56, the supposed aberrant specimen of G. leakeyi, described above, in its larger size and lower metaconid. It is quite likely that the development of the trigonid cusps on P_4 of Gymnurechinus is subject to much intraspecific variation.

The specimen of Pd_4 measures 3·6 × 2·0 mm, thus only slightly smaller than P_4, and it has a similar pattern, except that the trigonid cusps are more slender and acute but less high, and the notch between the paraconid and the protoconid is more open. The paraconid and metaconid are of equal height, as in the specimen of P_4. The height of the protoconid from the root bifurcation is 2·3 mm, compared with 3·1 in P_4.

M_1 is represented by an unworn specimen (Sgr 2013a.66) (figs. 6b, 6c) and a worn specimen associated with M_2 (Sgr 79.62). It is much larger than in G. leakeyi: the unworn specimen measures 5·1 × 3·4 mm and the worn specimen 5·1 × 3·6 mm, compared with 4·5 × 3·0 mm in a specimen of G. leakeyi. M_2 (fig. 6b) is also larger: 4·2 × 3·3 mm, compared with 3·8 × 2·7 mm in G. leakeyi. The molar patterns are the same in both species. The internal depth of the jaw of Sgr 79.62 at M_1 is 6·3 mm. Two fragments from Rusinga, Rs 145.51 (site unknown) and Rs 732.47 (site R3) contain poorly-preserved first molars which are indistinguishable from Sgr 79.62. The suggestion made previously (Butler, 1956b) that they might be referable to G. songhorensis rather than to G. camptolophus is strengthened. If adopted, it would follow that G. songhorensis coexisted with G. leakeyi at Rusinga site R3.

Amphechinus rusingensis Butler, 1956

Four additional mandibular specimens have been obtained:— From Rusinga: Rs 86.48, two associated rami, showing on the left side M_2–C and the bases of I_3 and I_2, and on the right side M_2–P_2 and the alveoli of M_3 and C, from site R105; Rs 409.56, a fragment with P_4, lacking most of the enamel, and the alveoli of M_1, from site R2. From Songhor: Sgr 1147.66, a fragment with right P_4 and the alveoli of the more anterior teeth; Sgr 1148.66, a fragment with left P_4.

There is some variation in the degree of crowding of P_2 (erroneously called P_3 by Butler, 1956b, p. 57). In Rs 86.48 it is separated from P_4 by a diastema 1·0 mm long; in the holotype P_2 is close to P_4 and its posterior root is displaced lingually; other specimens are intermediate.

Family *TENRECIDAE* Gray, 1821

Protenrec tricuspis Butler & Hopwood, 1957

Additional material.—Napak IV: right maxilla with P^3–M^3; left mandible with M_{1-3} and broken P_4 and P_3; both from trench 1c (1964).

Rusinga: Rs 551.49, incomplete right mandible with P_2–M_1 and the alveoli of M_2, from site R1.

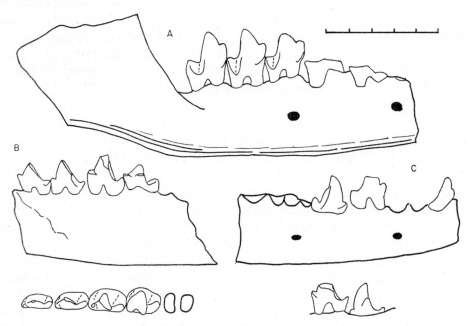

Fig. 7. *Protenrec tricuspis*. A, Napak specimen with M_3–M_1 and broken premolars. B, Rs 551.49, P_2–M_1, lingual and crown views. C, Sgr 3120c.66, fragment with P_4, P_3 and the canine, buccal view, and P_3–P_4, lingual view.

Songhor: Sgr 3120c.66, incomplete right mandible with the canine, P_3, P_4 and the alveoli of M_1 and M_2; Sgr 3120a.66, fragment of left mandible with M_1 and M_2; Sgr 2577.66, fragment with left P_3; Sgr 1911.66, fragment with alveoli, probably of M_1–P_3.

Lower jaw and teeth (fig. 7).—Of the lower dentition of this species, only the molars have been described hitherto. The new material provides information about the premolars. P_4 is longer and narrower than M_1. Its protoconid is a high acute cusp, seen unworn (incompletely erupted) in Sgr 3120c.66, higher than that of the unworn M_1 of the holotype or Sgr 3120a.66, but about equal in height to M_2. The metaconid is only about half the height of the protoconid. The paraconid is still lower (more so in Sgr 3120c.66 than in Rs 551.49) and anterior in position. The talonid resembles that of M_1 except that the basin on its lingual side is more reduced. P_3 is narrower than P_4 and slightly shorter. In Sgr 3120c.66 it resembles P_4 in pattern, but the tip of the protoconid has broken off; there is a posterolingual ridge representing the metaconid. In Sgr 2577.66 the tooth is more worn but probably similar in pattern. In Rs 551.49 the metaconid ridge is only slightly lingual in direction. P_2, preserved only in Rs 551.49, is a two-rooted tooth, shorter and narrower than P_3. The protoconid is tilted forwards to stand above the anterior root. There is no metaconid. The low talonid cusp is connected to the protoconid by a ridge which is slightly diverted towards the lingual side at the position where the metaconid would be expected. P_1 is absent. The canine, seen in Sgr 3120c.66, is a small, one-rooted tooth, not as high as the protoconid of P_4. It is tilted forwards; there is a rudimentary anterior cusp about halfway down the anterior ridge, and a cingulum-like talonid.

The symphysis extends back along the lower border of the mandible to below the posterior end of P_2. There are two mental foramina, below M_1 and P_2.

Maxilla and upper teeth. The only specimen showing the upper dentition is from Napak (fig. 8). It is referred to this species because by superposition of camera lucida drawings a satisfactory fit with the lower teeth is obtained. The teeth are moderately worn. M^1 and M^2 are of zalambdodont pattern, much wider than long, with the paracone placed somewhat lingually to the centre of the crown. Whether a metacone was present on the posterobuccal crest of the paracone cannot be definitely determined, but the shape of the worn crest is such that a metacone, smaller than that of *Potamogale*, may have been present close to the paracone. The protocone is about half the height of the paracone, and the ledge on which it stands occupies rather less than one-third of the width of the tooth, less than in *Potamogale* but more than in *Micropotamogale*. The protocone ledge is continued along the anterior margin as a cingulum which reaches the parastyle, but the posterior cingulum only reaches halfway across the tooth. The anterobuccal crest of the paracone connects with a stylocone, but the other buccal cusps have been eliminated by wear. M^2 is similar to M^1, but it is broader anteriorly and its posterobuccal lobe is somewhat reduced, rather more than in *Potamogale*. M^3, though wider than M^2, has undergone

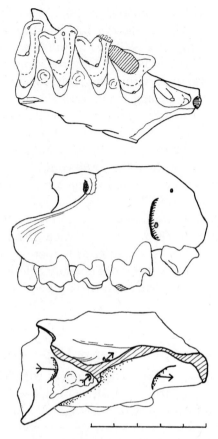

FIG. 8. *Protenrec tricuspis*, maxilla from Napak, with M³–P³, in palatal, lateral and dorsal views. In dorsal view, arrows pass through the infraorbital and lachrymal canals.

very considerable reduction of the posterobuccal lobe, and has a structure resembling *Potamogale*.

P⁴ is longer than M¹, due mainly to the forward projection of the parastyle region. The top of the crown has been broken off, and the arrangement of stylar cusps cannot be determined. The lingual lobe is stouter than in M¹, and the protocone is higher. It is probable that the paracone also was higher and stouter than on M¹. P³ is less molarized than in *Potamogale* : it has two roots, the posterior wider than the anterior, and is thus at an early stage in the development of the triangular shape. The protocone is represented only by an indistinct swelling at cingulum level. The centrally placed paracone has a posterior ridge, and the worn tooth shows an indication of a posterior buccal cusp. There is a low anterior parastyle. Immediately anterior to P³ is the posterior alveolus of P².

The short zygomatic process is fully preserved, showing that the zygomatic arch was incomplete, as in Recent Tenrecidae. The process arises above M^3. Its anterior face is hollowed out for the origin of the zygomaticus muscle. A strong ridge passes forward and upward from the process to form the lower edge of the orbit. The ridge continues on to the lachrymal bone to form the anterior edge of the lachrymal foramen, which therefore lies just within the orbit. The infraorbital canal is comparatively long: its posterior opening is opposite M^2 and the infraorbital foramen opens immediately anterior to P^4. There is a small foramen in the maxilla just anterior to the top of the infraorbital foramen. The lachrymal bone is separated from the maxilla by a distinct suture. It forms the anterior part of the orbital well, reaching the internal opening of the infraorbital canal. Its facial extension is small, and the foramen is situated near the anterior margin of the bone. More posteriorly on the lachrymal is a shallow pit. The lachrymal canal is quite short, only about 1 mm in length. In Recent Tenrecidae the infraorbital canal is much shortened to form a mere bridge, except in *Geogale aurita*, where it opens anterior to P^4 as in *Protenrec*. The ridge which marks the anterior border of the orbit is much weaker in the Recent forms, and it always fades out dorsally before reaching the level of the lachrymal foramen. The lachrymal foramen is absent in *Potamogale* and *Micropotamogale*.

Cf. *Erythrozootes chamerpes* Butler & Hopwood, 1957

Lower jaw and teeth.—Napak I: M21831, a fragment with left P_4–M_3 (fig. 9).

Songhor: Sgr 2491.66, a fragment from the right side with unworn M_3 and broken M_2 and M_1; Sgr 2186.66, a fragment with worn right M_3 and the base of the coronoid process.

This species was described on the basis of the anterior part of a skull from Koru. The lower teeth described here fit the upper teeth of the holotype.

The molars are stoutly built teeth resembling those of *Potamogale*. The protoconid is much the highest cusp. As seen in crown view, its ridges meet at an acute angle. The metaconid and paraconid are equal in height when unworn; the paraconid is a conical cusp at the anterolingual corner of the crown. An anterobuccal cingulum is present. The talonid, which is much lower and shorter than the trigonid, bears a single cusp, lower than any of the trigonid cusps. From the talonid cusp a ridge runs forward and slightly lingually towards the base of the metaconid, dividing a steep groove on its buccal side, for occlusion with the paracone, from the reduced talonid basin on its lingual side. The lingual edge of the basin is formed by a low ridge, weakly elevated to form a rudimentary entoconid. The three molars are similar in size, but the talonid of M_3 is somewhat longer and its cusp more elevated, though not as much as in *Microgale*. P_4 is a stout molariform tooth about the same length as M_1, but with a proportionately larger protoconid and shorter talonid. Its metaconid is about as high as on the molars, but it occupies a smaller area of the crown. The

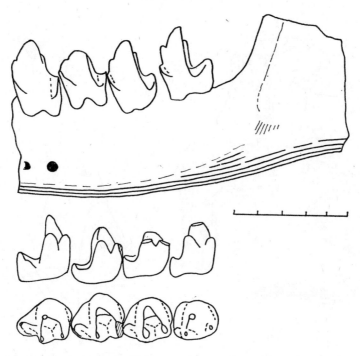

FIG. 9. Cf. *Erythrozootes chamerpes*, B.M. M21831, in outer view, and M$_3$–P$_4$ in lingual and crown views.

paraconid is smaller, a little lower than the metaconid but still conical and vertical; it is at the anterior end of the tooth, slightly to the lingual side.

The coronoid process, of which only the base is known, is strongly constructed and rises steeply from the tooth row. It is deeply excavated on its internal side but only weakly so externally. The horizontal ramus is stout. There are two mental foramina below P$_4$ in the Napak specimen, one below M$_1$ in Sgr 2491.66.

Cranium. M14316, a cranium from Koru (Maize Crib, surface), is referred to *Erythrozootes* because of its tenrecid characters and its size (fig. 10).

The broad somewhat flattened brain-case narrows rapidly to a tubular interorbital region. The sagittal crest is weak on the frontals but becomes better developed posteriorly (where it has been broken). The sutures of the skull roof have closed. The surface of the parietals is sculptured into a series of anastomosing ridges, rather less prominent than in *Gymnurechinus*; similar, but much fainter ridges are present in *Potamogale*, *Geogale aurita* and *Microgale dobsoni*. Several vascular foramina are present in the skull roof: two of large size high on the skull about 3 mm anterior to the nuchal crest, a smaller foramen near the squamosal-parietal suture, some lower in the squamosal, and a series of small foramina near the sagittal crest. In *Potamogale* the only conspicuous foramen is one near the upper margin of the

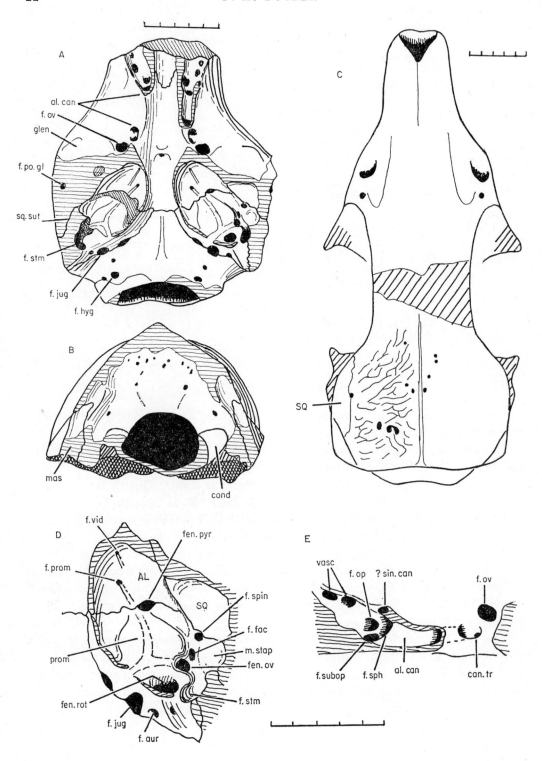

squamosal. The nuchal crest is strong, though its edge has flaked away. It becomes much weaker laterally, where it continues as the anterior lambdoid crest along the squamosal-mastoid suture, as in *Potamogale*. The mastoid exposure faces laterally, and the posterior lambdoid crest passes vertically down its posterior margin to the paroccipital process, which has been broken off.

The foramen magnum is low and transverse, its upper margin rising only slightly above the top of the condyles, as in *Potamogale*. There is a small posterior mastoid exposure, and more medially on the occipital a strong ridge forms the margin of the insertion of the m. obliquus capitis superior. There is no median crest on the occiput. A number of small foramina open below the nuchal crest.

The more projecting parts of the base of the skull have been broken away, fully exposing the large tympanic cavities, which converge anteriorly and are much closer together in the basisphenoid region than in *Potamogale* or *Micropotamogale*; they resemble *Geogale* and *Microgale* much more closely. The basioccipital is triangular, as the tympanic cavity approaches the condyle more than in *Potamogale* or *Micropotamogale*. The condyles approach each other in the middle line, as in Recent forms. The hypoglossal foramen is much smaller than in Recent Potamogalinae. More laterally a vascular foramen opens immediately anterior to the condyle; this has merged with the hypoglossal foramen in Recent Potamogalinae. A deep groove between the basioccipital and the periotic, partly roofed over, marks the course of the cavernous sinus. There is a foramen in the midline in the basisphenoid, as in other Tenrecidae, interpreted by McDowell (1958) as the opening for a persistent notochord; it is situated between the foramina ovalia as in most Tenrecidae. The posterior opening of the alisphenoid canal is situated shortly anteromedial to the foramen ovale, and its anterior opening is confluent with the sphenorbital foramen (foramen lacerum anterius). Unlike Recent Tenrecidae, the cranial wall of the alisphenoid canal is completely ossified. A small foramen opening just behind the posterior opening of the alisphenoid canal represents the transverse canal in the basisphenoid. There is no ectopterygoid process. The zygomatic process of the squamosal, the entoglenoid process, the postglenoid process and the region of the

FIG. 10. Cf. *Erythrozootes chamerpes*, B.M. M14316. A, B, ventral and posterior views of cranium. C, reconstruction of skull produced by combining the holotype with M14316 (unknown areas are cross hatched). D, left tympanic cavity. E, left orbital foramina. Abbreviations: AL, alisphenoid; al. can, alisphenoid canal; can.tr, transverse canal through basisphenoid; cond, occipital condyle; fen. ov, fenestra ovalis; fen. pyr, pyriform fenestra; fen. rot, fenestra rotunda; f. aur, auricular foramen; f. fac, facial foramen; f. hyg, hypoglossal foramen; f. jug, jugular foramen; f. op, optic foramen; f. ov, foramen ovale; f. po. gl, canal from postglenoid foramen; f. prom, foramen of promontorial artery; f. sph, sphenorbital foramen; f. spin, foramen spinosum; f. stm, stylomastoid foramen; f. subop, suboptic foramen; f. vid, opening of vidian canal; glen, part of glenoid surface; mas, mastoid exposure; m. stap, origin of stapedial muscle; prom, promontorium with grooves for carotid branches; ? sin. can, possible opening of sinus canal; SQ, squamosal; sq. sut, posterior suture of squamosal; vasc, vascular foramina.

external auditory meatus have all been flaked away, but it is possible to see the postglenoid canal in the broken surface.

The roof of the tympanic cavity is well exposed on both sides. The cochlear prominence is flatter than in *Gymnurechinus leakeyi*, and the groove for the pro-montorial branch of the carotid artery is more centrally placed, whereas in *Gymnurechinus* it is more lateral. A suture across the roof of the tympanic cavity anterior to the prominence separates the sphenoid and periotic bones. As in *Potamogale*, this suture is complete except for a small window immediately anterior to the prominence (equivalent to the pyriform fenestra of McDowell, 1958); in *Geogale*, *Microgale* and *Limnogale* the fenestra is much larger. As in Recent Tenrecidae and in *Gymnurechinus*, the anterior part of the chamber roofed by the sphenoid occupies about half the total length of the chamber. The promontorial artery pierced the sphenoid at a short distance anterior to the suture, as in *Potamogale*. Farther forward the opening of the vidian canal can be seen. The tympanic chamber extends back along the medial side of the prominence as far as the entry of the carotid, as in *Microgale* and *Geogale*; in the Potamogalinae this medial extension is not developed. As in other Tenrecidae the medial and posterior parts of the tympanic chamber were protected by flanges from the basisphenoid and periotic, but the extent of these flanges is unknown. The fenestra rotunda, the fenestra ovalis, the opening of the facial canal and the foramen spinosum (for the ramus superior of the stapedial artery) are much as in *Potamogale* and *Gymnurechinus*. More laterally is the depression for the stapedial muscle. The anterolateral part of the tympanic chamber is roofed by the squamosal, forming a broad but shallow epitympanic recess, resembling that of *Potamogale* in which, however, it is more extensive. The medial suture of the squamosal is oblique as in *Potamogale*, whereas in *Microgale* it is more longitudinal. In *Gymnurechinus* the contribution of the squamosal to the tympanic roof is much smaller. The medial edge of the squamosal is developed in *Erythrozootes* into a sharp ridge, passing from the foramen spinosum to the anterior end of the chamber. The groove on the medial side of this ridge presumably indicates the course of the ramus inferior of the stapedial artery.

A small part of the orbital wall is visible. The lateral margin of the sphenorbital foramen is continued as a strong ridge passing forwards and upwards. A small foramen, probably for the lateral cerebral sinus, opens in the edge of this ridge, and more anteriorly two vascular foramina open downwards below the ridge. Opening into the medial wall of the sphenorbital foramen are two foramina, separated by a narrow bridge of bone; the upper one is the optic, the lower the suboptic foramen. The optic foramen is comparatively large, as in *Potamogale*; in other living Tenrecidae, including *Micropotamogale*, it is much more reduced and in many cases it is represented only by a groove in the medial wall of the sphenorbital canal. The suboptic foramen is very small in *Potamogale*, though well developed in *Micropotamogale*, *Limnogale* and probably other genera.

Relationships of the Miocene Tenrecidae

At the present time the Tenrecidae have an interesting discontinuous distribution. On the African continent occur *Potamogale velox* from the forest regions of the Congo Basin, *Micropotamogale lamottei* from West Africa (Mount Nimba) and *M. ruwenzorii* from the lower slopes of Mount Ruwenzori. (Guth, *et al.*, 1959, 1960; Verheyen, 1961a,b). These comprise the subfamily Potamogalinae. The remainder of the family are found in Madagascar, where they show a degree of adaptive radiation wide enough to suggest that the family has existed on the island for a long time, perhaps since the Miocene or earlier. The presence of three genera of Tenrecidae in the Miocene of Africa is therefore of interest.

Protenrec is known from the upper dentition as far forward as P^3, the lower dentition as far forward as the canine, most of the lower jaw (except the extremities) and the anterior part of the orbit. It agrees with *Potamogale* in many respects, but most of the resemblances are shared with at least some of the Malagasy tenrecids and are probably primitive characters of the family: presence of an anterior cingulum on the upper molars and an anterobuccal cingulum on the lower molars (these cingula are absent in *Micropotamogale*, at least in *M. ruwenzorii*); M^3 wider than M^2 (M^3 is reduced in some specimens of *Potamogale* and in *Micropotamogale*); three premolars; canines small (in *Protenrec* the last two characters are known only in the lower jaw, but it may be presumed that the upper teeth were similarly differentiated); M_3 similar in size to M_2 (smaller in *Micropotamogale*); posterior mental foramen present. *Protenrec* differs from the Malagasy forms but resembles *Potamogale* in the relatively large protocone of the upper molars and of the talonid basin of the lower molars; in these respects it is intermediate between *Potamogale* and *Micropotamogale*. If the angulation of the posterior ridge of the paracone indicates a metacone, this would be a further resemblance to *Potamogale*. However, if the zalambdodont molars of Tenrecidae have been derived from a dilambdodont condition one would expect the presence of the protocone and metacone in early stages of evolution, and the resemblances in molar pattern between *Protenrec* and *Potamogale* may merely indicate the retention of primitive characters in the living genus.

In the length of the infraorbital canal *Protenrec* is more primitive than any other known tenrecid, being approached only by *Geogale*. The prelachrymal crest is also a primitive character, found also in *Solenodon* and Erinaceidae. *Potamogale* and *Micropotamogale* have lost the lachrymal foramen, perhaps an aquatic adaptation, for it is much reduced in *Limnogale*. In *Protenrec* P^3 and P^4 are at a lower level of molarization than in living Potamogalinae. P^3 has only two roots, whereas in all living tenrecids it has three roots, except in *Geogale* where it is reduced. P_2 in *Protenrec* has two roots and an anterior basal cusp, as in *Microgale* and *Oryzorictes*, whereas in the Potamogalinae and also in *Limnogale* it has one root and lacks the basal cusp.

Thus *Protenrec* appears to be near the primitive stock from which both the

Potamogalinae and the Malagasy tenrecids have been derived. It is specialized in two respects: (1) the posterior buccal lobe of M^2 is reduced, as in *Micropotamogale*, *Limnogale* and *Geogale aurita*; (2) there is a marked contrast in size between P^3 and P^4, foreshadowing the condition of *Geogale*.

Erythrozootes is known from the upper dentition, the lower dentition as far forward as P_4, and most of the skull. It approaches *Potamogale* in (1) the relatively well-developed protocone and talonid basin; (2) the degree of ossification of the tympanic roof; (3) the comparatively large optic foramen; (4) the triangular, three-rooted condition of P^3 (shared with the Malagasy forms except *Geogale*); (5) the upper border of the foramen magnum, which does not rise above the level of the condyles; (6) the oblique medial suture of the squamosal in the tympanic roof. Of these characters, (1)–(3) are probably primitive features lost in the Malagasy forms but retained in the Potamogalinae; (4) could be either primitive or a parallel development; (5) and (6) are of uncertain significance. Against these resemblances to *Potamogale* must be placed a number of differences: (1) P^2 is absent in *Erythrozootes*, as in *Geogale aletris*; (2) the cranial roof is very rugose, with numerous foramina; (3) the lachrymal foramen is present; (4) the infraorbital canal is longer, opening above P^4 instead of M^1; (5) the hypoglossal foramen is smaller; (6) the suboptic foramen is well developed, as in *Micropotamogale* and the Oryzorictinae; (7) the cranial wall of the alisphenoid canal is fully ossified; (8) the upper canine is one-rooted, as in *Oryzorictes* and some specimens of *Microgale* (two-rooted in Potamogalinae and other Oryzorictinae); (9) the tympanic cavity is larger, with a medial expansion, as in *Microgale* aud *Geogale*. Of these differences from *Potamogale*, (1) and (2) are due to specialization of *Erythrozootes*, in (3)–(7) *Erythrozootes* retains a more primitive condition; (8) is of uncertain significance, and (9) is possibly indicative of a relationship of *Erythrozootes* with the Malagasy forms.

Thus neither *Erythrozootes* nor *Protenrec* falls into any of the living subfamilies, and it is proposed to erect a new subfamily to include them:—

Subfamily **Protenrecinae,** nov.

Diagnosis: Primitive Tenrecidae with the following characters: protocone well differentiated on the upper molars; talonid basin occupying about half of the talonid on the lower molars; M^3 wider than M^2; P^4 much longer than M^1; upper canine small, not as large as I^1 (unknown in *Protenrec*, where the lower canine is known to be small); M_3 similar to M_2 in size, its talonid only moderately enlarged; lachrymal foramen present; infraorbital canal opens anteriorly to or above P^4, long or of moderate length; anterior mental foramen below P_3 (unknown in *Erythrozootes*), the posterior one below M_1 or P_4.

Included genera: *Protenrec* Hopwood and Butler, *Erythrozootes* Hopwood and Butler, both from the Miocene of East Africa.

The two genera, in so far as they can be compared, show significant differences such as the absence of P² in *Erythrozootes*, indicative of a long period of evolutionary diversification on the African continent prior to the Miocene.

Geogale aletris, the third type of Miocene tenrecid, is known only from one rather poorly-preserved specimen, showing the upper dentition and the facial part of the skull. The molars are not well enough preserved to determine the size of the protocone. In the enlargement and separation of the first incisors, the reduction in size of P³ and the long infraorbital canal it resembles *G. aurita* now living in Madagascar. There are, however, certain differences from the living species: P² is missing in the Miocene form, the posterobuccal lobe of M² is less reduced, the infraorbital foramen opens above P⁴ instead of anterior to P⁴, and the palate extends further back posteriorly to M³. Some of these differences would exclude *G. aletris* from the direct ancestry of *G. aurita*, and they raise the possibility that the similarities are due to parallel evolution.

Geogale aurita is usually placed in the subfamily Oryzorictinae, with *Microgale* and *Oryzorictes*, but it differs rather markedly from these in its somewhat soricid-like dentition, and also in the retention of primitive characters such as the relatively long infraorbital canal, the retention of the posterior mental foramen below M_1 and the lack of fusion of the tibia and fibula; in the last two characters it resembles the Tenrecinae. It is probably best placed in a separate subfamily, Geogalinae Trouessart (1879, p. 275). If it is assumed that the Malagasy Tenrecidae are monophyletic, *Geogale* would have branched off at an early stage. On the other hand, if the resemblance between *G. aurita* and *G. aletris* is indicative of a real relationship, *Geogale* must have invaded Madagascar independently, for its dentition is too specialized for it to have given rise to the other Malagasy tenrecids.

Van Valen (1967) transferred the Tenrecidae, Solenodontidae and Chrysochloridae from the order Insectivora to the order Deltatheridia, of which they comprised a suborder Zalambdodonta. His reason for doing this appears to be a belief in their possible derivation from the Palaeoryctidae, based on similarities in the molar pattern to *Palaeoryctes* (see Van Valen, 1966, pp. 104–105). On the same ground Matthew (1913) placed *Palaeoryctes* near the ancestry of the Tenrecidae. However, McDowell (1958), as the result of a restudy of *Palaeoryctes*, found a number of important differences from Tenrecidae in the base of the skull, particularly in the ear region: there is a groove medial to the petrosal leading to a carotid foramen as in *Leptictis* (*Ictops*) (Butler, 1956) and *Didelphodus* (Van Valen, 1966); the promontorium is not grooved for the promontorial artery (though Van Valen, 1966, identifies a foramen in the anteromedial part of the tympanic chamber as the promontorial foramen); there is no tympanic process of the basisphenoid, and the fenestra rotunda is fully exposed, instead of being partly covered over by a wing of the periotic (the bulla in *Palaeoryctes* was formed by a tympanic or entotympanic). Except for the possible loss of the promontorial artery, these features are all primitive characters lost or modified in the Tenrecidae and other Lipotyphla. Although the occipital region of *Palaeoryctes* is poorly known, it is probable that the post-tympanic part of its skull was very short, and that the mastoid exposure faced backwards as in *Leptictis* and *Zalambdalestes* rather than laterally as in Lipotyphla. Other primitive features are the interorbital constriction (shared with Erinaceidae), the ossification of the medial wall of the alisphenoid canal (shared with Erinaceidae and *Erythrozootes*), and the relatively small extension of the tympanic chamber anteriorly to the periotic (shared with *Solenodon*). The infraorbital canal (seen in Univ. Kansas 7748) is fairly long, opening anteriorly to M^1: in the related genus *Puercolestes* (Reynolds, 1936) it opens anteriorly to P^4 as in *Protenrec*. The lachrymal foramen in the University of Kansas specimen of *Palaeoryctes* opens just within the orbit, as in *Protenrec*. These resemblances to *Protenrec* merely emphasize the primitive nature of the Miocene genus; they do not indicate any special relationship. *Palaeoryctes* appears to have an essentially primitive eutherian skull, whereas the skulls of Tenrecidae share numerous non-primitive characters with other Lipotyphla.

The upper molars of *Palaeoryctes* resemble those of *Potamogale* in being much broader than long, and in that the metacone is smaller than the paracone, closely applied to it and standing more buccally. However, the paracone and metacone are both very high cusps, much higher than the buccal stylar cusps, and the protocone occupies a larger area on the crown. In the lower molars the trigonid is much more elevated, especially the protoconid, the paraconid is less distinct, and the talonid, though low, has a normal structure. The molars appear to have functioned by a nearly vertical movement of the jaw, as in other Palaeoryctidae, whereas in Tenrecidae the jaw movement has a large transverse component as in Soricidae (Mills, 1966). The dental resemblances between *Palaeoryctes* and Tenrecidae are not sufficiently close to over-ride the numerous differences in skull structure, and the Tenrecidae should therefore be left in the Lipotyphla.

Butselia, from the L. Oligocene of Belgium, has been interpreted as a primitive tenrecoid by Quinet and Misonne (1965) and as a deltatheridian related to *Didelphodus* by Van Valen (1967). It differs from *Didelphodus* in the closer approximation of the metacone to the paracone, the presence of a well-developed hypocone-cingulum and apparently also in the more lingual position of the conules. It is later than *Didelphodus*, which died out in N. America in the M. Eocene. *Butselia* resembles Tenrecidae in the differentiation of the buccal stylar cusps, including a strongly-developed stylocone (anticone in Vandebroek's nomenclature) and in the approximation of the metacone to the paracone. However, the metacone is directly posterior to the paracone and subequal to it in height, and the lingual part of the tooth has a structure found in Leptictidae and Pantolestidae. In the associated lower molars the talonid is larger and more normal in structure than in *Potamogale*. While it is not impossible that the tenrecid molar pattern could have been derived from one resembling *Butselia*, there is no evidence that it has done so, and *Butselia* is best left *insertae sedis*. Nothing is known of its skull.

Solenodon, from the West Indies, has usually in the past been placed in the super-family Tenrecoidea. However, McDowell (1958) showed that in the characters of its skull *Solenodon* closely resembles *Nesophontes*, which has dilambdodont molars. He concluded that *Solenodon* and *Nesophontes* are more closely related to the Soricidae than to the Tenrecidae, and that the zalambdodonty of *Solenodon* had been acquired independently. *Protenrec* resembles *Solenodon* and *Nesophontes* in the prelachrymal crest, a primitive character found in erinaceoids, but it differs from them in the long infraorbital canal, in which it is probably more primitive. *Erythrozootes* differs from both *Solenodon* and *Nesophontes* in a number of characters of the basicranium: the medial wall of the alisphenoid canal is ossified; the basisphenoid possesses a tympanic process; the tympanic cavity is much larger, extending forward well beyond the periotic; its roof is much more ossified: there is a broad epitympanic recess, roofed by the squamosal. Inter-radicular crests appear to be absent from the molars of both *Protenrec* and *Erythrozootes* as in Recent tenrecids, but present in *Solenodon* and *Nesophontes*. Thus the Miocene forms do not significantly close the gap between the Solenodontidae and the Tenrecidae; on the contrary, they show an approach in some characters towards the Erinaceidae.

The author does not discuss here *Apternodus*, which according to McDowell is not a lipotyphlan, and in the opinion of Van Valen (1966) may be descended from Palaeoryctidae. Van Valen (1967) treats the Apternodontinae as a subfamily of the Tenrecidae, which seems unjustified on present evidence.

The Tenrecidae appear to be a family of African origin, derived from an unknown group of primitive Lipotyphla, and not closely related to any other family of living insectivores except possibly the Chrysochloridae. Until more is known about the early history of the Lipotyphla it is not possible to be more precise than this.

Family *CHRYSOCHLORIDAE* Mivart, 1868

Prochrysochloris miocaenicus Butler & Hopwood

Additional material, from Songhor: Sgr 359.62, anterior part of a skull and associated lower jaw, now separated; Sgr 598.66, a similar specimen with part of the lower jaw and the lower cheek teeth in occlusion (fig. 11).

Description.—The two additional specimens are preserved in much the same way as the holotype and the paratype. Sgr 598.66 shows more of the tip of the snout than the holotype, and posteriorly it is broken away to expose part of the internasal septum. Sgr 359.62 lacks the tip of the snout, but posteriorly it shows a little more of the base of the cranium. The teeth of Sgr 598.66 are little worn, and P_3 on the left side is incompletely erupted. In this specimen the internasal suture and the premaxilla-maxilla suture are closed, but the maxilla-nasal and maxilla-frontal

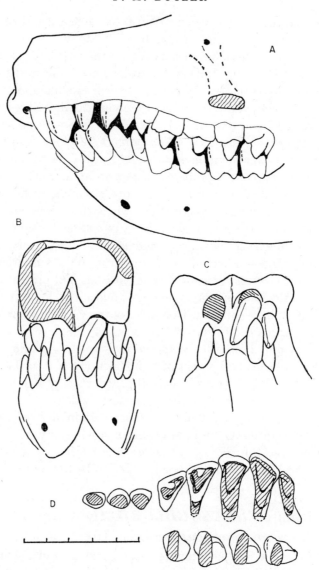

FIG. 11. *Prochrysochloris miocaenicus.* A, reconstitution of dentition based mainly on
Sgr 598.66, with supplementary details from Sgr 359.62. B, C, Sgr 598.66, anterior and
ventral views of anterior end of specimen. D, Sgr 359.62, upper dentition (lacking I^1
and I^2) and P$_4$–M$_3$, crown view. Details of the unworn cusp pattern of the upper cheek
teeth have been added from Sgr 598.66.

sutures are still open, as in the holotype. In Sgr 359.62, which has worn teeth, all
the sutures are closed.

Sgr 598.66 shows that the snout is broadened in the premaxillary region, the
premaxillae being very thick laterally to the external nares, as in Recent chryso-

chlorids. The lateral thickenings, however, do not project forwards beyond the median septum, unlike Recent forms. The ventral part of the median septum is ossified to form a stout median process, which is continued ventrally as a keel between the alveoli of the first incisors. In the degree of the forward projection of the process beyond the incisors *Prochrysochloris* resembles *Chrysospalax trevelyani*, which it also resembles in the shape of the process.

The lateral wall of the infraorbital canal is not preserved in either specimen. The lachrymal foramen is visible in both: it is extraorbital, and seems to have been situated above the dorsal end of the infraorbital bridge as in most living forms, and not posterior to it as in *Chrysospalax*. The zygomatic process is represented only by its root, best seen on the right side of Sgr 359.62, where it extends from the posterior part of M^1 to the posterior part of M^2. In most Recent species it is somewhat farther forward, opposite M^1; *Chrysospalax* is peculiar in its relatively long zygomatic root, which extends along the whole length of M^1 and M^2.

The cribriform plate is well exposed in Sgr 359.62, and behind it the floor of the brain-case can be seen for about 3 mm. There is no optic foramen. The presphenoid is very thick and filled with spongy bone which extends into the lateral wall of the sphenorbital canal.

The mandible is represented in both specimens by the anterior parts of the two rami. The rami are stout and probably meet at a rather large angle, though in both specimens there has been some displacement at the symphysis. The symphysis is short and deep, extending back to about the level of P_2 as in the living species. There are two mental foramina, below P_2 and M_1; in living forms they vary in number and position, but are rarely as far forward as P_2 (*Amblysomus*) or as far back as M_1 (*Chrysospalax*).

Sgr 359.62 shows all the upper teeth from I^3 backwards, but in a worn condition. In Sgr 598.66 I^1 and I^2 are present on the left side, I^3 and the canine on the right side; P^4–M^3 are well preserved on both sides but are overlapped by the lower molars in occlusion. I^1 is the highest incisor; it originally pointed somewhat backwards and medially, though pushed further backwards *post mortem*. It is simple in structure, with a high, acute cusp, apparently with mesial and posterobuccal ridges (incompletely exposed), as in the living forms. A ridge also passes down the anterior surface, which is more evenly rounded in the living forms. I^2 is less elevated and less acute than I^1, but it is longer anteroposteriorly. The anterior ridge is more prominent, and is marked off by a vertical groove on the buccal side, as in *Chrysochloris asiatica* and *Amblysomus corriae*. There is a posterior angulation at cingulum level, as in some of the living species. I^3 is a much smaller tooth than I^2, and apparently blunter (though its tip is slightly hidden by the lower teeth); its anterior ridge is displaced towards the buccal side, and the vertical buccal groove is distinct. The canine resembles I^3 in size and shape; its root, fully exposed in Sgr 359.62, measures 1·8 mm, apparently about the same length as the buccal roots of the molars. The lingual surfaces of the anterior teeth are not visible in Sgr 598.66, but in Sgr 359.62 I^3 and the canine are seen to be longer than broad, with a lingual convexity situated posteriorly to the middle of the tooth.

P^2, which is known only from the worn specimen, is a one-rooted tooth similar in size to the canine, but differing in the greater size of the lingual convexity, so that the breadth of the tooth is about equal to the length. In the Recent chrysochlorids it is more molariform. P^3 of Sgr 359.62 resembles the corresponding tooth of the paratype; it is partly molariform, with a small, cingulum-like protocone and a high posterior buccal cusp; there is no parastyle. In Recent forms P^3 is fully molariform. P^4 is well preserved in Sgr 598.66, but its protocone is not exposed. It is molariform. Sgr 359.62 shows that the highest cusp, the paracone, is placed about half-way across the tooth, and lingually there is a low but well-developed protocone. The paracone is V-shaped, ridges passing from its tip along the anterior and posterior margins of the crown. The buccal margin is occupied by an elevated ridge, somewhat indistinctly differentiated into cusps: there is an anterior buccal cusp, incompletely separated from the parastyle, and a slightly higher posterior buccal cusp. Sgr 598.66 shows that the molars have a similar pattern, but on M^2 the posterior buccal cusp is less high and on M^3 it is absent. The protocones of the molars appear to be higher than on P^4, and on M^3 the protocone can be seen to be V-shaped and not very much lower than the paracone. The molariform teeth of Recent Chrysochloridae resemble *Prochrysochloris* in the structure of the buccal margin, but the protocone is more reduced; moreover, the emphasis is farther forward in the series, M^3 being more reduced or completely lost.

All the lower teeth are visible in buccal view in Sgr 598.66. The mandibular rami in the molar region have broken away, leaving the molars in place, interlocked between the upper molars; only the anterior root of left M_1 is still embedded in the jaw. The molar crowns are visible in Sgr 359.62 but in a worn condition. The largest lower incisor is I_2, which occludes between I^1 and I^2. I_1 is a small narrow tooth, placed near the symphysis mesially to I_2; it is only about three-fifths of the height of I_2. I_2 is inclined forward at an angle of about $45°$ to the tooth row. It is bluntly pointed, convex buccally, with an anterior ridge and an apparently sharper posterior ridge. There is no evidence of a posterior basal cusp, but the posterior part of I_2 is overlapped buccally by I_3 and not fully visible; a posterior basal cusp is present in Recent forms. I_3 is much smaller than I_2, about the height of I_1. It is blunter at the tip than I_2, and it possesses a faint vertical groove on the buccal side, separating the anterior ridge from the general buccal surface. The lower canine is a little larger than I_3, though much smaller than I_2. It resembles I_3 in pattern, but the buccal groove is very faint.

P_2 is a one-rooted tooth, about the same length as the canine but less elevated: it possesses a posterior basal cusp. In Recent forms P_2 is more molariform, usually with some indication of a metaconid. P_3 is about the same height as P_2, but its main cusp is more vertical and more convex buccally; the anterior ridge is turned more lingually, and the posterior basal cusp has the appearance of a simple talonid. In Sgr 598.66 this tooth is in process of eruption on the right side, though fully in place on the left. P_3 seems to be the last lower tooth to erupt in *Chlorotalpa stuhlmanni* and perhaps in other living species. P_4 is a very much larger tooth, about twice as high as P_3. The protoconid is by far the highest cusp, the paraconid and metaconid being only about half as high; there is a faint anterobuccal cingulum and a small talonid; the root appears to be undivided. The molars have two roots and are nearly equal in size. The talonid is much narrower than the trigonid and it occupies rather less than half the length of the tooth, except on M_3 where it is somewhat longer. The trigonid is much higher than the talonid; the metaconid is directly lingual to the protoconid, and in the worn condition the two cusps are of similar height, though the protoconid was probably relatively higher when unworn; the paraconid seems to be reduced and less lingual than the metaconid. A weak anterobuccal cingulum is present. Details of the talonid cannot be determined, but there was at least one well-developed cusp, reaching on M_3 about half the height of the metaconid. The lower molars differ from those of Recent chrysochlorids in the greater length and width of the talonid, the greater development of the talonid cusp especially on M_3, the clearer separation of the roots, the presence of an anterobuccal cingulum, and the greater equality in size of P_4–M_3.

Discussion.—The additional information now available about *Prochrysochloris* confirms the conclusions previously reached as to its relationships. It is more primitive than any living species in the dentition, particularly the upper and lower molar patterns, the lower level of molarization of P_2^2 and P_3^3, and the stronger development of the posterior molars. At the same time it is a typical chrysochlorid in characters of the skull, including those of the tip of the snout. Its upper molars, though primitive in their relatively large protocones, resemble those of Recent chrysochlorids in the development of the buccal cusps.

Prochrysochloris shows that the Chrysochloridae and the Tenrecidae were well differentiated by Miocene times, and it throws little light on the possibility of a common ancestry of the two families. Compared with *Protenrec*, *Prochrysochloris* is more specialized in its molar patterns: higher trigonid, less differentiated paraconid, weaker anterobuccal cingulum on the lower molars, poor differentiation of a parastyle on the upper molars. The infraorbital canal of *Prochrysochloris* is reduced to a bridge, and its lachrymal foramen is extra-orbital. It had already developed the characteristic specializations of the premaxillae and the cancellous bone of the base of the skull found in living chrysochlorids, and it had advanced beyond *Erythrozootes* in the loss of a separate optic foramen. Almost the only known resemblance to Tenrecidae that might be of phylogenetic significance is the zalambdodont molar pattern, but this may well be due to parallel evolution: reduction of the protocone, at least, has taken place in both families since the Miocene. In our present state of ignorance it seems best not to place the Chrysochloridae in the superfamily Tenrecoidea, but instead to retain the Chrysochloroidea as a separate superfamily.

Order CHIROPTERA Blumenbach, 1779
Family *EMBALLONURIDAE* Dobson, 1875

Genus and Species Indeterminate

An unnumbered distal end of a left humerus, from Rusinga site R3, belongs to this family (fig. 12). It much resembles *Taphozous* (incl. *Saccolaimus*) and agrees in size with *T. longimanus*. The joint surface is in line with the shaft of the humerus, as in Emballonuridae, Vespertilionidae and Molossidae. The lateral keel of the capitellum is slightly concave when seen from the anterior side, its proximal end projecting as an epicondyle rather more markedly than in *Taphozous*. On the medial side, the epitrochlea projects somewhat more medially than in *Taphozous* and its styloid

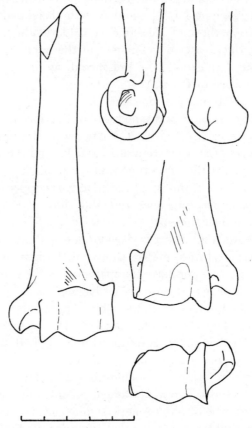

FIG. 12. Emballonurid humerus from Rusinga, in anterior, lateral, medial posterior and distal views.

process is shorter; in these respects the fossil resembles *Coleura seychellensis*. The maximum diameter of the distal end of the humerus is 5·0 mm.

Taphozous has already been recorded from Koru (Butler and Hopwood, 1957) on the basis of a skull fragment. The Emballonuridae range back in Europe to *Vespertiliavus* from the Phosphorites of Quercy (U. Eocene or L. Oligocene) (Revilliod, 1920, p. 95). The humerus of this genus differs from the Rusinga specimen in the weaker development of the lateral epicondyle, but resembles it in the rudimentary styloid process.

Family *HIPPOSIDERIDAE* Miller, 1907
(?) *Hipposideros* sp.

Sgr 149.62, from Songhor, is the distal end of a right humerus which closely resembles *Hipposideros diadema* except for its smaller size. (fig. 13) The maximum diameter of

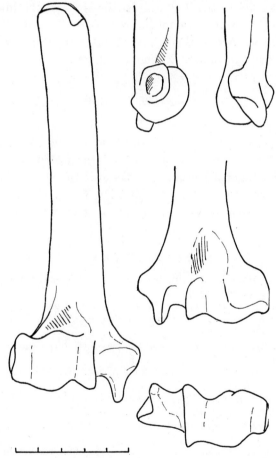

FIG. 13. (?) *Hipposideros* sp., Sgr 149.62. Humerus in anterior, lateral, medial, posterior and distal views.

the distal end is 5·7 mm. The shaft of the bone is nearly circular in cross section in its middle portion, but becomes broader and flatter distally. It is straight in anterior view but curved towards the anterior side. The distal joint surface is displaced towards the lateral side as in *Hipposideros* and *Rhinolophus*. The posterior depression for the ulnar sesamoid and the anterior depression above the capitellum for the radius are similar to *Hipposideros* and less deep than in *Palaeophyllophora* (Revilliod, 1917, p. 27). The capitellum is spherically rounded, and not keeled as in *Rhinolophus*. The width of the epitrochlea, measured from the trochlear keel medially, is more than half the width of the articular surface. The epitrochlea has a medial extension which is truncate at the tip and a strongly-developed styloid process, flattened medio-laterally, which projects beyond the trochlear keel as in *Hipposideros* and *Pseudo-rhinolophus*. In *Pseudorhinolophus* the styloid process is longer. The styloid process of the Songhor specimen is situated nearer to the trochlea than in *Rhinolophus*, and it is directed parallel to the long axis of the shaft of the humerus, whereas in *Rhinolophus* it diverges towards the medial side.

This is the oldest member of the Hipposideridae to be recognized from Africa. The family extends back to the M. Eocene in Europe.

ACKNOWLEDGMENTS

I wish to express by best thanks to Dr. L. S. B. Leakey and Dr. W. W. Bishop for sending me their specimens for study, and also to Mrs. S. C. Savage and other members of the staff of the British Museum (Natural History) for much practical help with fossil and Recent material. Thanks are also due to Dr. J. C. McKenna, Dr. L. Van Valen, Professor B. Patterson and Dr. W. A. Clemens for their kindness in assisting me with comparative material during a visit to the United States.

REFERENCES

BISHOP, W. W. 1964. More fossil Primates and other Miocene mammals from North-east Uganda. *Nature, Lond.* **203**: 1327–1331.

BISHOP, W. W. 1965. The later Tertiary in East Africa—Volcanics, sediments, and faunal inventory. In *Background to Evolution in Africa*, ed. W. W. BISHOP and J. D. CLARK. Pp. 31–56. University Chicago Press.

BUTLER, P. M. 1956a. The skull of *Ictops* and the classification of the Insectivora. *Proc. zool. Soc., Lond.* **126**: 453–481

—— 1956b. Erinaceidae from the Miocene of East Africa. *Br. Mus. (nat. Hist.) Lond., Fossil Mammals of Africa* No. 11: 1–75.

BUTLER, P. M. and HOPWOOD, A. T. 1957. Insectivora and Chiroptera from the Miocene rocks of Kenya Colony. *Br. Mus. (nat. Hist.) Lond., Fossil Mammals of Africa* No. 13: 1–35.

FILHOL, H. 1891. Étude sur les mammifères fossiles de Sansan. *Ann. Sci. géol., Paris*, **21**: 1–319.

GUTH, C., HEIM DE BALSAC, H. and LAMOTTE, M. 1959/1960. Recherches sur la morphologie de *Micropotamogale lamottei* et l'évolution des Potamogalinae. *Mammalia* **23**: 423–447, **24**: 190–217.

JAMES, G. T. 1963. Paleontology and nonmarine stratigraphy of the Cuyama Valley badlands, California. Part I. Geology, faunal interpretation, and systematic descriptions of Chiroptera, Insectivora, and Rodentia. *Univ. Calif. Publs. geol. Sci.* **45**: 1–154.

KENT, P. E. 1945. The Miocene beds of Kavirondo, Kenya. *Q. Jl. geol. Soc. Lond.* **100**: 85–118.

LE GROS CLARK, W. E. and LEAKEY, L. S. B. 1951. The Miocene Hominoidea of East Africa. *Br. Mus. (nat. Hist.) Lond., Fossil Mammals of Africa* No. 1: 1–117.

McDOWELL, S. B. 1958. The Greater Antillean insectivores. *Bull. Am. Mus. nat. Hist.* **115**: 113–124.

MATTHEW, W. D. 1913. A zalambdodont insectivore from the Middle Eocene. *Bull. Am. Mus nat. Hist.* **32**: 307–314.

MATTHEW, W. D. 1918. A revision of the Lower Eocene Wasatch and Wind River faunas. Part V.—Insectivora (continued), Glires, Edentata. *Bull. Am. Mus. nat. Hist.* **38**: 565–657.

MILLS, J. R. E. 1966. The functional occlusion of the teeth of Insectivora. *J. Linn. Soc.* **46**: 1–25.

PATTERSON, B. 1965. The fossil elephant shrews (family Macroscelididae). *Bull. Mus. comp. Zool. Har.* **133**: 295–335.

QUINET, G. E. and MISONNE, X. 1965. Les insectivores zalambdodontes de l'oligocène inférieur belge. *Bull. Inst. R. Sci. nat. Belg.* **41** (19): 1–15.

REVILLIOD, P. 1917–1922. Contribution à l'étude des chiroptères des terrains tertiares. *Abh. schweiz paläont. Ges.* **42**: 1–60, **44**: 61–130, **45**: 131–195.

REYNOLDS, T. E. 1936. Two new insectivores from the Lower Paleocene of New Mexico. *J. Paleont. Chicago*, **10**: 202–209.

SCHLOSSER, M. 1911. Beiträge zur Kenntnis der oligozaenen Landsäugetiere aus dem Fayum, Aegypten. *Beitr. Paläont. Geol. Ost.-Ung.* **24**: 73–88.

SEEMANN, I. 1938. Die Insektenfresser, Fledermäuse und Nager aus der obermiozänen Braunkohle von Viehhausen bei Regensburg. *Palaeontographica* **89A**: 1–56.

SHACKLETON, R. M. 1951. A contribution to the geology of the Kavirondo Rift valley. *Q. Jl. geol. Soc. Lond.* **106**: 345–392.

THENIUS, E. 1949. Zur Revision der Insektivoren des steirischen Tertiärs. *Sber. öst. Akad. Wiss.* Abt I. **158**: 671–693.

TROUESSART, E.-L. 1879. Catalogue des mammifères vivant et fossiles. *Rev. Zool. Paris*, ser. 3, **7**: 219–285.

VAN COUVERING, J. A. and MILLER, J. A. 1969. Miocene stratigraphy and age determination, Rusinga Island, Kenya. *Nature, Lond.* **221**: 628–632.

VAN VALEN, L. 1966. Deltatheridia, a new order of mammals. *Bull. Am. Mus. nat. Hist.* **132**: 1–126.

—— 1967. New Paleocene insectivores and insectivore classification. *Bull Am. Mus. nat. Hist.* **135**: 217–284.

VERHEYEN, W. N. 1961. Recherches anatomiques sur *Micropotamogale ruwenzorii*. *Bull. Soc. R. Zool. Anvers*. Nos. 21, 22.

VILLALTA, J. F. and CRUSAFONT-PAIRÓ, M. 1944. Nuevos insectivoros del mioceno continental des Vallés-Panadés. *Notas Commun. Inst. geol. Miner. Esp.* **12**: 39–65.

VIRET, J. 1940. Étude sur quelques Erinacéidés fossiles (suite). Genres *Plesiosorex, Lanthanotherium*. *Trav. Lab. Géol. Univ. Lyon* **39**: 33–65.

WEBB, S. D. 1961. The first American record of *Lantanotherium* Filhol. *J. Paleont.* **35**: 1085–1087.

WHITWORTH, T. 1954. The Miocene hyracoids of East Africa. *Br. Mus. (nat. Hist.) Lond. Fossil Mammals of Africa* No. 7: 1–58.

MIOCENE CERCOPITHECOIDEA AND OREOPITHECOIDEA FROM THE MIOCENE OF EAST AFRICA

G. H. R. VON KOENIGSWALD

Senckenberg Museum, Frankfurt am Main, Germany

INTRODUCTION

The Cercopithecoidea, the monkeys, comprise a most flourishing group of higher Primates, inhabiting Africa, the tropical parts of Asia (reaching Japan) and even occur in Gibraltar on the soil of Europe. Two subfamilies are generally recognized, Cercopithecinae and Colobinae. The first subfamily, according to Buetter-Janusch (1963), contains 6 genera and 37 species, the second 7 genera and 23 species. The subspeciation is astonishing: of *Cercopithecus aethiops* 23 subspecies are recognized, of *Cobobus badius* 21.

In spite of this great diversity of genera, species and subspecies very little is known about the history of the group. There is a fair number of Pleistocene forms, partly belonging to extinct genera, but the Tertiary record is very poor. *Mesopithecus* from the Pontian of Greece is the only species, of which enough of the post-cranial skeleton has been preserved to compose a more or less complete skeleton (Gaudry, 1882). Most species are only known by dentitions. From the Miocene of Africa till now only two finds have been published (MacInnes, 1943; Hooijer, 1963), and we are very fortunate that Dr. L. S. B. Leakey and his collaborators have now assembled some materials in East Africa, which form the subject of this study. We are greatly

4

obliged to Dr. Leakey, for entrusting this valuable material to us, as well as for his great patience in waiting for the manuscript. The finds allow some definite conclusions but the material is still very incomplete, and it seems that in the meantime no further finds can be added to the list. While in the fauna of today there are many more monkeys than anthropoids, according to a list published in 1951, Le Gros Clark and Leakey had on record 214 remains of anthropoids against only 13 of Cercopithecids.

Dr. Leakey's material not only adds two new species to the list, but gives us a definite answer as to the origin of the cercopithecine molar pattern. All forms up to beginning of the Pliocene have a highly specialized, bilophodont pattern; in that they are higher evolved than the Hominoidea, including man, with which they share the same tooth formula with only two premolars. It is just this bilophodont type of dentition which makes it very easy to recognize a cercopithecoid, but makes it so difficult to erect and distinguish species. As in the Hominoids it has been taken for granted, that the additional fourth cusp of the upper molar should be a true hypocone, even if earlier stages were not yet known. Hürzeler, in his new definition of *Oreopithecus*, has challenged this statement and caused a vivid discussion, without contributing to a solution. A proper explanation of this problem is more than a purely academic dispute as it involves the relationship between Cercopithecoidea and Hominoidea. Fortunately the material collected by Dr. Leakey allows us to answer that question.

A surprise is the presence of an *Oreopithecus* from Moboko, indicating an intramiocene specialization of the European species.

CERCOPITHECOIDEA

The first Cercopithecinae in Europe appear in the Lower Pliocene of Eastern Europe. *Mesopithecus* is not rare at the classical site of Pikermi; Gaudry already had some 40 specimens at his disposal, and was even able to reconstruct the entire skeleton. Monkeys are, however, absent from the Lower Pliocene of India where they first appear in the Middle Pliocene of the Dhok Pathan, and in China they do not occur probably before the Upper Pliocene. In the immense material available from the Chinese Pontian in the drug-stores, where mammalian teeth are sold as dragon teeth (l'ung tse) we never observed a single specimen. The centre of evolution of the Cercopithecidae therefore must be somewhere in Africa, which is also evident from the materials described in this paper.

From the " Upper Miocene " of Oran, Algeria, Arambourg (1959) has described a new species, *Macaca flandrini*. The fauna, containing *Hipparion* and *Hyaena*, is typical Pontian and, according to our standards, should better be regarded as Lower Pliocene, in spite of occurring in beds intercalated with a marine mollusc fauna of Sarmatian (Upper Miocene) type. But this is not the place to quarrel about the

Mio-Pliocene boundary. The dentitions figured suggest two species rather than one. A slightly smaller form with well-developed *lobus tertius* in the last lower molar (Pl. **1**, fig. 3) and more bunodont upper molars (Pl. **1**, fig. 1) is contrasted by a more robust dentition with a much shorter *lobus* (fig. 5) and nearly lophodont upper molar (fig. 4).

In 1943 MacInnes for the first time described finds from the Lower Miocene of Kenya, which he compares with *Mesopithecus*? From Ongoliba in the Congo, Hooijer (1963) reported an elongated third lower molar with well-developed *lobus tertius*, " he referred to " cf. *Macaca* c.q. *Mesopithecus* sp. This tooth is a typical cercopithecoid molar, that is all that can be said. But it is the only find outside East Africa, indicating a wider distribution of these monkeys in the African Miocene.

Superfamily **CERCOPITHECOIDEA** Simpson 1931
Family *CERCOPITHECIDAE*
Subfamily **Victoriapithecinae** nom. nov.
Genus *Victoriapithecus* n.g.

Diagnosis: Primitive Cercopithecidae; bilophodonty in the upper molars not yet fully developed, original trigon still recognizable, and by that different from Cercopithecinae and Colobinae. True hypocone as in the Hominoidea and Oreopithecoidea.

Victoriapithecus macinnesi n.sp.

1943 *Mesopithecus* Wagner? MacInnes, Notes etc. pp. 148–151; Pl. 23, fig. 1.
Type specimen: lower jaw from Rusinga, figured by MacInnes.

Material:

Lower jaw

Fragment right mandible, P_4–M_3, R.S. (type)
fragment left mandible, M_2–M_3
fragment left mandible, M_2–M_3 MB 49
left M_3
symphysis of lower jaw without teeth. OMBO
8 lower canines.

Upper jaw

Right anterior premolar, MB 6–51
fragment of first or second molar, MB 8–51
first or second molar, 206–59.
second or third molar right,
worn second or first molar, NBA
third right molar, MB 46–49.

The material available consists of some fragments of lower jaws; the upper dentition is only represented by isolated teeth. We distinguish two forms, which can be separated according to size and the pattern of the upper molars. But there is a fair chance that more than two species are present; even in recent material it seems sometimes impossible to distinguish definite species on account of isolated teeth alone (Sampel). As the East African anthropoids, the cercopithecoids show a puzzling variability.

The best specimens of the whole collection is still the lower jaw, figured 1943 by MacInnes; this is made the type specimen of a new species, called *macinnesi* in honour of its discoverer.

It is the fragment of a right ramus, containing the three molars and the last premolar. Depth of the ramus at M_3 16 mm, thickness at M_2 7·5 mm, length of P_4–M_3 25 mm.

The premolar is set obliquely and is bicuspid; the cusps are united by a crest. The molars are elongated and of typical cercopithecine pattern; first and second molar with four cusps, arranged in pairs, third molar with simple third lobe, the tip of which is set in one line with protoconid and hypoconid. Third lobe less than one third of the crown. Labial cusps more worn than lingual, with triangular wearing facet.

The upper molars mentioned by MacInnes are probably our figs. 4, 5 and 9. None of them has the quadrangular outline, typical of the present day cercopithecinae; in all of them the posterior part is constricted. In fig. 5a the hypocone is not yet fully developed, and the original trigon dominant; the specimen of fig. 9 shows a crest, descending from the protocone in the direction of the metacone, but not reaching its tip. In the specimen fig. 4 there are no crests, the hypocone is fully developed, but the metacone reduced. The material is insufficient to make sure what might be individual variation and what might indicate eventual different species.

We must mention here a worn first or second molar, well worn with rounded cusps, and in spite of attrition still showing the clear traces of a *crista obliqua* and a well-developed hypocone; outer cingulum present. The specimen, fig. 12, for that reason was first united with the second species, but by the absence of an anterior ridge between para- and protocone might well belong to *macinnesi*.

Upper and lower canines are indistinguishable from those of living cercopithecoids. The upper canine of the male is laterally compressed, with a marked groove running up to the apex of the tooth. There are two sizes of upper canines, of which the smaller is attributed to this species. The length of the canine fig. 15 is 25 mm, greatest thickness 6·2 mm, while the height of the crown is 11 mm. In the lower canines there is a large cingulum at the lingual side, ascending on both sides to the top of the canine; in the lower canine, fig. 18 and referred to the smaller species, the apex is broken off. The height of the crown is 12 mm, the oral root a maximum diameter of 7 mm and a minimum diameter of 5 mm.

Victoriapithecus leakeyi n.sp.

Slightly larger than *macinnesi*, and referred to the same genus.

Type specimen: an upper second molar from Kiboko, figured in Pl. 1, fig. 8.

Material:

Lower jaw

M_3 MB. (fig. 7) M_3 MB and M_3 MB 2–51
fragment lower jaw, MB 54 P_4–M_2
fragment lower jaw, MB 11/51 M_1 M_2
fragment lower jaw, M_1 KBA
lower molar, M_1 MB 216–49
lower molar, M_2 MB 48–42
lower premolar, P_3 MB 224–49
2 lower canines P_3
VB P_3
2 lower canines
lower molar M_2 MB 49 7·7–7·0.

Upper jaw

First upper premolar (referred specimen).
second upper molar type, KBA
first upper molar, MBA
third upper molar, MBA
2 upper M^2
One large upper canine female; referred specimen
6 referred upper canines.
skeletal material: fragment of ulna.

The larger species can best be distinguished on account of the upper molars. They approach already more or less the quadrangular type common in cercopithecidae, but the anterior portion of the crown is still broader than the posterior part. There is a marked crest connecting the tip of the protocone with the tip of the metacone, the *crista obliqua* indicating the old trigon. Thus the fourth cusp is a normal hypocone.

While there is not yet a ridge developed between hypocone and metacone, the formation of the cusps is typical cercopithecoid, especially the sharp ridge which in the unworn molar fig. 8 begins at the anterior border, and runs over the tip of the paracone further over the tip of the metacone to end at the posterior border. Paracone and metacone are already directly connected by a sharp ridge, closing of a large *fovea anterior*. A thin ridge descends from the tip of the protocone down into the central valley. An outer cingulum is well developed.

The same sharp crests, connecting meta- and entoconid and posto- and hypoconid are characteristic of an unworn third molar, which is larger than in *macinnesi*. No transversal ridges are developed. The third lobe is large, length about one third of the tooth. The unworn specimen is figured in fig. 6, and worn specimen with about the same dimensions in fig. 7, worn first and second molars in figs. 10 and 11.

According to its dimensions an upper premolar is referred to this species, fig. 13. All three roots are preserved. The crown is narrow, both cusps are pointed and the outer is distinctly higher than the lingual.

There seem to be two types of larger upper canines. A short one, worn out with a triangular crown and a triangular root, must be regarded as belonging to a female specimen, fig. 14,—length 29 mm, thickness of root 8·9 mm—while an elongated, thin, sharply pointed type must represent a male specimen; fig. 16, length of the tooth 28·5 mm, height of the crown 14 mm, thickness of root 8·5 mm.

There is a fragment of the upper extremity of the ulna, apparently belonging to the larger species, fig. 1. The greatest length of the specimen is 56 mm; at the lower end it is flattened, transversal diameter is 8 mm, the thickness 5 mm.

The upward prolongation is typical cercopithecoid and agrees well with a specimen of *Cercopithecus aethiops* at our disposal, the latter being a little smaller and more slender. The diameter of the *incisura* is 8·8 mm; the *processus coronoidens* is accentuated.

In summary it can be said that the fragment, smaller than that of *Limnopithecus macinnesi*, shows the cercopithecoid specializations already developed and is very different from the same bone of *Mesopithecus*, where the upward extension of the olecranon is much lower.

CONCLUSIONS

The dentition of the living cercopithecoid monkeys is highly specialized, in fact it is more specialized than that of the Hominoidea, including Man and the great apes. In the molars of the monkeys the original cusps are united by transversal crests, making them " bilophodont ". Lophodont dentitions, apparently adapted to soft food, are to be found in many animals. Of living forms we might mention the *Tapir* among the *Perissodactyla* and *Macropus* among the Marsupialia. Extinct are *Listriodon*, belonging to the Suidae, *Pyrotherium* of the Amblypoda, and *Dinotherium* and *Zygolophodon* of the Proboscidea.

By Lower Pliocene time, when the first Cercopithecoidea appear in Europe and Asia—*Mesopithecus pentelicus* in Pikermi and *Macacus sivalensis* in the Siwaliks of India—the bilophodont dentition is already fully established, and there is no clue about their origin.

It has, however, always been assumed that in the upper molar the anterior ridge para- and protocone are connected, and in the posterior ridge meta- and hypocone respectively (Gregory, 1922, p. 298).

When Hürzeler (1949) redescribed the dentition of *Oreopithecus*, a primate which by some authors has wrongly been regarded as a " forme de passage " between the Cercopithecoidea and the Pongidae, a form which possesses a typical hypocone, he discussed at the same time the fourth cusp of the *Cercopithecus* upper molar.

There are according to Hürzeler, three possibilities:—

(1) the cusp is a true hypocone, derived from the cingulum;

(2) we are dealing with a " pseudohypocone " derived from the protocone;

(3) the fourth cusp is an enlarged *metaconulus*.

Without giving any proof, Hürzeler from the very beginning suggested, that the cusp in question was not a true hypocone. As an enlarged *metaconulus* does not form a main cusp in any of the lower primates, that possibility—realized in some artiodactyla—can be ruled out right away. The question hypocone or pseudo-hypocone has been discussed by Kälin (1955, p. 10). Also by the present author (1955), who, however, has been corrected. " Since 1951 I have, on the contrary, been inclined to take it for an interodistally shifted, and fortified metaconule " (Hürzeler, 1958, p. 30). Since 1951 no evidence has been published to prove this revolutionary statement, and Kälin more recently has taken up the discussion " hypocone or pseudo-hypocone " again (Kälin, 1962, p. 35, fig. 1).

The question of the interpretation of that disputed cusp is of much more than of purely academic importance, it determines, to an important degree, the phylogenetic possibilities of the Cercopithecoidea. As already mentioned, the Hominoidea and Oreopithecoidea possess a true hypocone. If the cusp in question of the *Cercopithecoidea* is a hypocone, then a closer relationship to the Hominoidea might be possible, if it is a pseudo-hypocone, then both groups are separated from the very beginning, and must have originated from different lower primates, still in the possession of the original trigonodont pattern.

The molars of the Pliocene, Pleistocene or recent cercopithecoid monkeys furnish no clue to solve the question. But primitive traits often survive in the deciduous dentition—as the paraconid in the last lower deciduous molar of gorilla—and on that base Remane (1951) has studied the dentitions of *Cercopithecus* and *Colobus*. Among his results

(1) the anterior upper deciduous molar of the Cercopithecoidea shows a variation from nearly trigonodont to bilophodont pattern;

(2) the *crista transversa* posterior in the upper molars has been formed by two different parts. The labial part is identical with the labiodistal part of the *crista obliqua*, the lingual part consists of a ridge originating from the hypocone.

Remane's observations have been confirmed by Lampel (teste Kälin, 1962); also by unpublished observations by the present writer, who had a large material, fossil and subrecent, at his disposal. We found a clear formation of what can only be the original trigon not only in a number of deciduous molars, but also in a permanent molar with an underdeveloped talon. Hürzeler (1958, p. 31) has attacked

Remane violently, calling " such studies . . . irrelevant for the understanding of the primates and neglect them altogether ". Here it might be added, that the last lower deciduous molar, a tooth which in higher primates generally resembles greatly the first permanent molar, in *Oreopithecus* looks very different from that tooth and shows a pattern which demonstrates " die isolierte Stellung von *Oreopithecus* " (Remane, 1955, p. 488).

We might close this dispute with the remark, that the study of the deciduous dentition has given no clue that the hypocone of the Cercopithecoidea could be anything else but a " true " hypocone.

The evolution of the cercopithecoid monkeys before they arrived in the Pontian in Eurasia, must have taken place in Africa. It was, however, not before 1943, that MacInnes described the lower jaw of a typical cercopithecoid monkey he called " *Mesopithecus* " from the (Middle) Miocene of Kenya in East Africa.

While the detailed description is to be found elsewhere in this paper we might here draw attention to the fact, that in *Victoriapithecus* the *crista transversa* in the upper molars, is completely preserved. The canines are already fully cercopithecoid.

So while the lower molars of *Victoriapithecus* are typically such of a cercopithecoid, the upper molars are not yet fully bilophodont. The molar cusps are already compressed with a transversal crest, and the crista obliqua connecting the tip of the broad protocone with the metacone leaves no doubt, that the fourth cusp can only be a " true " hypocone, even without a cingulum visible. A connection metacone-hypocone is not yet indicated. *Oreopithecus* has rounder cusps, a cingulum and inflated metaconulus.

The discovery of *Victoriapithecus* brings to an end all speculations about origin and age of the cercopithecoid dentition. Gregory's opinion (1922, p. 299) regarding *Apidium* and *Oreopithecus* as cercopithecoid forerunners, has only historical interest. Also Kälin (1962) seems to be inclined to regard the bilophodont dentition as something very ancient and compares the pattern of *Gesneropithecus* and *Alsaticopithecus*, two apparently lemuroid primates from the Eocene of Europe, to those of the Cercopithecoidea.

The complete bilophodonty of the cercopithecoid monkey has been acquired during the Miocene. They form by that an astonishing parallel to a group of specialized Suidae, the Listriodontinae. *Listriodon lockharti* from the Middle Miocene of Europe was still partly bunodont; *Listriodon splendens* from the Upper Miocene was so completely lophodont, that in Steinheim Fraas has even described it as " *Tapirus suevicus* ". In India the bunodont form is *Listriodon guptai* from the Kambial, the lophodont form *Listriodon pentopotamiae* from the Chinji, which is hardly distinguishable from the European *splendens*. Here we witness in two regions that, apparently by the pressure within the flourishing group of the Suidae, the strong competition and the need for specialization in connection with the conquering of new biotopes the bilophodont dentition is developed within a short period.

Another example, of which the time factor however is not known, can be found among the Primates themselves. Among the Pleistocene Lemurs of Madagascar we find on one side the *Megaladapis edwardsi*, with still triconodont upper molars, a hypocone not yet developed. On the other side we have the Archaeolemurinae, with bilophodont molars in the upper and the lower jaw. Abel (1962) has given good pictures of both *Archaeolemur* and *Bradylemur*. An upper jaw of *Archaeolemur* in the author's collection shows a strong cingulum at the first and second molars, proof of the presence of a " true " hypocone in these forms, of which the geological history unfortunately is still unknown.

The pre-Miocene Cercopithecoidea must have had a dentition not so different from that of early pongids. The primary connection metaconid–protoconid in the lower molars probably was the only crest already present; the surface must have been smooth with no wrinkles, and the hypoconulid weak or absent and in a medial position. In the upper molar, teste *Victoriapithecus*, the original *crista obliqua* was still present, but no metaconulus.

The molars of Hylobates might serve as a model; already Bolk (1914) has published a *Symphalangus* with nearly bilophodont teeth. Also Hürzeler (1954, fig. 6d) has figured an upper molar of *Hylobatus lenciscus* from the Leiden collection, which shows similar conditions. Also *Parapithecus* from the Oligocene of Egypt, recently excellently redescribed by Kälin (1961), might perhaps belong to the same category. The Hylobatidae, possess in the trigonid part of the lower molar only a single crest (" die wohl als hintere Trigonidleiste zu bezeichnen ist. Von der vorderen Trigonidleiste war nie irgend eine solche Spur nachzuweisen "). Remane (1922, p. 77), while in the pongids there are two crests, " vordere und hintere Trigonidleiste " of Remane. Because of the presence of only a single crest Remane has regarded *Propliopithecus* and *Parapithecus* as primitive Hylobatidae. In the Cercopithecidae, as far as we can judge from the present-day conditions, in geologically older forms conditions might be similar. In a second lower deciduous molar of a modern *Hylobates* we found only a single crest, no paraconid, small cusps and an incipient hypoconulid in median position. This in striking difference to the same tooth in *Pliopithecus* (Hürzeler, 1954; fig. 24a), where we observe a paraconid and two crests; the conditions here are closer to the pongids than to *Hylobates*.

We might conclude by saying that the bilophodont dentition of the present-day Cercopithecidae is modern and that, from an odontological point of view, a closer affinity to the Hylobatidae might be possible. The few fossils known are difficult to interpret and not conclusive, but perhaps the degree of relationship between the Cercopithecidae and the Hylobatidae on one and between the latter and the pongidae on the other side could perhaps be determined by modern biochemical analysis, giving us a clearer and more modern picture of the phylogenetic structure of the family tree of the higher primates.

A first beginning we might find in the comparative study of the chromosomes.

As Klinger and co-workers reported (1963, p. 239) : " the karyotype of the gibbon is completely different and resembles more that of the specialized Cercopithecoid monkeys of the genera *Macaca*, *Papio* and *Cercocebus*. Perhaps in view of this data a taxonomic revision of the group, particularly in relation to the position of the gibbon, is called for ".

In this connection it is of great importance that recently, Ankel (1965) could show that *Pliopithecus* of the European Miocene, because of size and dentition, till now generally regarded as a member of the Hylobatinae, has had a large hollow *canalis sacralis* in its sacrum, which allows the conclusion that this form was still in the possession of a long tail. This surprising observation certainly excludes *Pliopithecus* from the Hylobatinae s.str. ; this form probably might be a last survivor of the original group, which gave rise to the bilophodont cercopithecoidea which mostly have tails, and are by that characteristic different from the Hominoidea too.

Our observations of the East African Cercopithecoids also show, that all speculations about " cercopithecoid dentitions " of Oligocene forms have no real foundation ; this concerns especially *Apidium* and *Moeripithecus* from Egypt.

On finishing this paper Mr. L. E. M. de Boer, in classifying fossil and recent teeth of cercopithecoid monkeys from the East Indies, has found in the collection of the Zoological Institute of the University of Utrecht two skulls of *Macaca fascicularis* from Banka (nr. 39), and Kuala Lumpur respectively, which show in the upper molars a very well-developed cingulum, comparable to the same structure in the Hominoidea. Especially in the specimen from Banka in both second molars the cingulum entirely surrounded the protocone; in the first molars nothing was visible, the third molars still in the crypt. So at last he had found what neither Remane nor the author could observe: a typical cingulum in a cercopithecoid monkey. The survival of this structure within this group apparently is very rare, but it certainly is not an abnormality.

OREOPITHECOIDEA

Oreopithecus from the Lower Pliocene (Pontian) of Northern Italy and Southern Russia has sometimes been regarded as a cercopithecidae, sometimes as an anthropoid, even as an early Hominine. Cercopithecoides and Hominoides can be separated on account of their dentition, and it is evident that the peculiar cusp pattern we observe in *Oreopithecus* is not intermediate between both types of dentition, that *Oreopithecus* is not merely a " forme de passage ", but represents a type of its own, which is also evident from a complete skeleton discovered by J. Hürzeler, but not yet completely described.

Till now *Oreopithecus bambolii* was the only species known ; it is apparently a terminal form, no other finds are known from post-Pontian sediments. *Apidium* from the Lower Oligocene of the Fayûm has been formerly regarded as a possible

TABLE I

Dimensions of teeth referred to *Victoriapithecus*

| | Upper Dentition | | | | |
	P³	P⁴	M¹	M²	M³
MB 51/6 worn	6·2–6·0–4·0				
206 left?			6·5–7·5		
MB 46–49 right (fig. 4)					5·4–7·3
M²? right (fig. 5)			7·0–7·2		
KBA completely worn ? right (fig. 13)	5·4–6·6–4·5			6·5–6·8	
KBA left type (fig. 8)				8·0–8·0	
MBA right (fig. 9)					5·9–6·9
288			7·9–8·5 7·4–8·0		
MBA right fig. 12			7·0–7·3		

P³ 3 roots, anterior cingulum strongly developed; bicuspid with high pointed tubercles (referred specimen).

| | Lower Dentition | | | | |
	P₃	P₄	M₁	M₂	M₃
Mand. RS. fig. 1 type (r)		4·0–4·0	5·6–5·0	6·8–5·9	7·8–5·5
M₂, M₃ MB 49 fig. 3 (l)				6·8–6·0	7·0–5·4
M₃ green number (r) weathered					7·7–5·5
M₂ M₃ fig. 2 (l)				7·7–6·1	9·5–5·8
KB fig. 6 (r)					9·5–7·0
OMBO (r)					9·4–6·3
MB 49–409 (r)					8·9–6·5
MB 4–51 (r)					8·4–6·2
MB 49 (l)					10·4–7·0
MB 41 fig. 7 (r)					10·3–7·4
MB 2–51 (l)					10·0–6·4
MB 54 fig. 10 (r)		6·0–4·5	6·0–6·0	7·8–7·5	
MB 1151 fig. 11 (r)			6·2–5·6	7·5–7·0	
KBA (r)				7·5–6·5	
MB 49 (l)				7·7–7·0	
MB 216–49 (r)			7·0–7·2	?	
M 48–42 (l)				7·0–7·3	
VB	7·0–5·5				

Remarks

Symphysis ± 15 mm, breadth between the canines 7 mm, canine left length 6·5 mm, breadth 3·2 mm.

MB 41 M₃   MB 51/11 M₁ M₂ M₁–M₂–M₃ $\dfrac{13·5 + 10·3 =}{= 23·8}$ large species.

Mand. RS   M₁–M₃   = 20·7 small species

P₃ MB 224–49 damaged and worn off, most of the enamel gone.

Presumably not yet monocuspid.

ancestor (Simons, 1960), but newly discovered materials have shown that this form was still in the possession of three premolars.

As in other categories of higher primates an African centre of origin could also be expected for the Oreopithecoidea. Among the materials from Maboko and kindly sent to me by Dr. Leakey, there is a Miocene member of this group, indicated by a single molar. The new form is generically better separated from the Pontian form; the new species we will call *clarki* in honour of our old friend Sir Wilfrid Le Gros Clark, who has contributed so much to our knowledge of the African primates. Leakey has noted the presence of " Oreopithecus " recently at Fort Ternan (Leakey, 1968).

Mabokopithecus clarki n.g.n.sp.
(Pl. **1**, fig. 20).

The only specimen (MB 51–1) and therefore the type, is a last left lower molar. The tooth is practically unworn, but part of the hypoconulid is missing. The trigonid part is wider than the talonid portion, a primitive character at the labial side is slightly curved. The metaconid is the most prominent cusp, and connected with the protoconid by a simple, slightly interrupted crest. Behind the crest at about the middle of the tooth, is a distinct tubercle (called unfortunately " mesoconid " by Hürzeler), which is very typical for this primate.

A cingulum descends from the middle of the anterior part, that can be followed around the lingual side and apparently merged with the hypoconulid. The hypoconid is well developed, with a slight longitudinal crest. Between metaconid and endoconid a well separated tuberculum intermedium.

While the general impression of this rather elongated molar suggests an *Oreopithecus*, the position of the talonid cusps is entirely different. In the latter form hypoconid and entoconid are opposite to each other, so that worn molars could easily have been mistaken for teeth of Cercopithecidae (Hürzeler, 1949; figs. 3, 4). In our specimen from Maboko the hypoconid is in juxtaposition, resembling the conditions in the anthropoids, and as in the latter endoconid and hypoconulid are joined by a crest, closing of a *fovea posterior*, which has been destroyed when the tooth was damaged; whether or not a *tuberculum sextum* was present, cannot be decided.

It is because of the different arrangements of the talonid cusps that we propose a new genus, *Mabokopithecus*. The length of the tooth is 10·7 mm, the transverse diameter is 7·3 mm. The tooth, therefore, is smaller than in *Oreopithecus*, where the measurements are 14·1 and 10·2 mm respectively (I.Z.F. 4580).

Mabokopithecus is more primitive than *Oreopithecus* in the following characteristics:

> trigonid wider than talonid;
> external cingulum better developed;
> trigonid crest better developed;

fovea posterior closed by crest;
smaller size.

The new genus differs fundamentally by the presence of
" anthropoid cusp arrangement of the talonid part ".

What has been already discussed elsewhere in this chapter that the anthropoid cusp pattern is more primitive and has gradually changed into the bilophodont pattern of the Cercopithecoidea, seems to be apparently true also for the Oreopithecoidea. The lower molars, in the position of the cusps and the elongation of the third molar, *Oreopithecus*, approaches but not fulfills cercopithecoid conditions. The reason for that probably has been a parallel and independent adaptation for soft food, probably leaves.

It is interesting to note that at about the same time (Pontian) also among the Dryopithecinae there is a close, but not complete approach towards bilophodonty. The teeth from the Snabian fissure fillings are proof of it. In *Dryopithecus meficus* we find in the upper molars an apparently constant connection between hypocone and metacone (Branco, 1898, Pl. 1, figs. 1, 2), which is still recognizable when the tooth is worn; in the lower molars the crests closing fovea anterior and posterior respectively, are about parallel (idem, Pl. 2, figs. 1, 6).

We see no reason why *Mabokopithecus* from the Miocene could not be ancestral to *Oreopithecus* from the Pliocene.

REFERENCES

ABEL, O. 1931. *Die Stellung des Menschen im Rahmen der Wirbeltiere.* pp. 1–398. Fischer, Jena.

ANKEL, F. 1965. Der Canalis sacralis als Indicator für die Länge der Caudal region der Primaten. *Folia Primat.* nr. **3** 263–276.

ARAMBOURG, C. 1959. Vertebres continentaux du Miocene Supérieure de l'Afrique du Nord. *Publ. Serv. Geol. Algerie, Pal.* **4:** 1–159.

BOLK, L. 1914. Odontologische Studien. Jena.

BUETTNER-JANUSCH, J. 1963. *Evolutionary and Genetic Biology of Primates.* Academic Press, New York.

GREGORY, W. K. 1922. *The Origin and Evolution of the Human Dentition.* pp. 1–548. Williams, Baltimore.

HOOIJER, D. A. 1963. Miocene Mammalia of Congo. *Kon. Mus. Tervuren, reeks* 8, **46:** 1–77.

HÜRZELER, J. 1949. Neubeschreibung von *Oreopithecus bambolii* GERVAIS. *Schweiz. Pal. Abh.* 66, pp. 1–20.

KÄLIN, J. 1955. Zur Systematik und evolutiven Deutung der höheren Primaten. *Experientia* **11:** 1–17.

—— 1961. Sur Les Primates de l'Oligocene Inferieur d'Egypte. *Ann. Pal.* **48:** 1–48.

—— 1962. Ueber *Moeripithecus markgrafi* SCHLOSSER und die phylogenetischen Vorstufen der Bilophodontie der Cercopithecoidea. *Bibl. Primat.* **1:** 32–42.

KLINGER, H. P., HAMERTON, J. L. and MUTTON, D. 1963. *The Chromosomes of the Hominoidea.* Viking Fund Publ. **37,** pp. 235–242.

LEAKEY, L. S. B. 1968. Upper Miocene Primates from Kenya. *Nature, Lond.,* **218,** 527.

LE GROS CLARK, W. and LEAKEY, L. S. B. 1951. The Miocene Hominoidea of East Africa. *British Museum, Foss. Mammals of Africa,* **1:** 1–117.

MACINNES, D. G. 1943. Notes on the East African Primates. *J. East Africa and Uganda Nat. Hist. Soc.* **xvii:** 141–181.

REMANE, A. 1951. Die Entstehung der Bilophodontie bei den Cercopithecidae. *Anat. Ann.* **98:** 161–165.

SIMONS, E. L. 1960. *Apidium* and *Oreopithecus*. *Nature, Lond.* **186:** 824–826.

PLATE 1

Victoriapithecus macinnesi n.g.n.sp.

FIG. 1. Right mandible RS with P_4–M_3. Small form. A, buccal; B, lingual; C, occlusal view; D, occlusal view, twice enlarged. Specimen figured by MacInnes 1943; Pl. 24, fig. 1. Type specimen.

FIG. 2. Fragment of left mandible with M_2–M_3. Large form. A, buccal; B, occlusal view; C, occlusal view twice enlarged.

FIG. 3. Fragment of left mandible MB–49 with M_2–M_3. Small form. Occlusal view.

FIG. 4. Right upper third molar MB–46, occlusal view. 4A, twice enlarged.

FIG. 5. Right upper (first or second?) molar; A, twice enlarged.

FIG. 12. First right upper molar, M.B.A. Referred specimen. Occlusal view. A, twice enlarged.

FIG. 15. Right upper canine, male type. Lingual view.

FIG. 18. Right lower canine, lingual and labial view, Referred specimen.

Victoriapithecus leakeyi n.sp.

FIG. 6. Right third lower molar KB; unworn. Occlusal view. A, twice enlarged.

FIG. 7. Right third lower molar MB–41; worn. Occlusal view; A, twice enlarged.

FIG. 8. Left second upper molar K.B.A.; unerupted, occlusal view. A, twice natural size. Type specimen.

FIG. 9. Right third upper molar M.B.A. Occlusal view; A, twice enlarged.

FIG. 10. Fragment of right mandible, MB–54, with P_4–M_2. Occlusal view.

FIG. 11. Fragment of left mandible, MB. 11–51, with M_1–M_2. Occlusal view.

FIG. 13. Upper third premolar; referred specimen. A, buccal, B; anterior; C, occlusal view.

FIG. 14. Left upper canine, female type.

FIG. 16. Right upper canine, male type. Lingual view. B, twice enlarged.

FIG. 17. Right lower canine. B, lingual; D, labial view; idem A and C twice enlarged.

FIG. 19. Left proximal fragment of ulna, lateral views. Referred specimen.

Mabokopithecus clarki n.g.n.sp.

FIG. 20. Left third molar MB. 1–51. Occlusal view, A, twice enlarged. Type specimen.

If not otherwise indicated the figures are natural size.

PLATE I

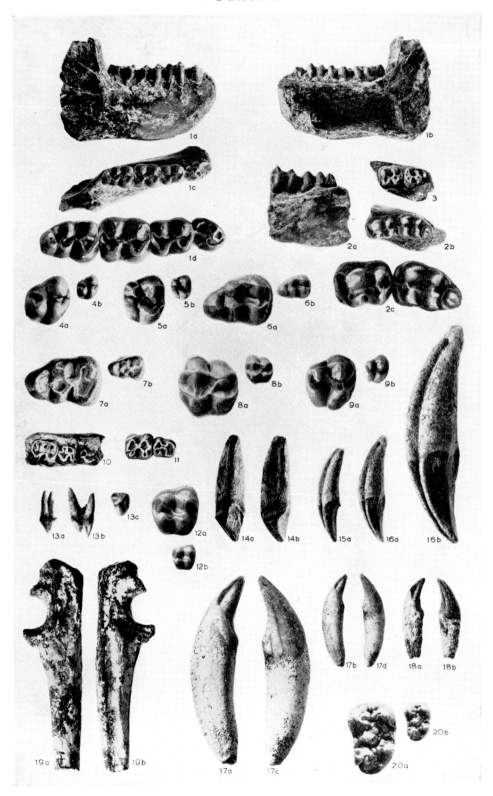

NEW CERCOPITHECIDAE FROM THE CHEMERON BEDS OF LAKE BARINGO, KENYA

R. E. F. LEAKEY

Kenya National Museum, Nairobi

INTRODUCTION

The two specimens described in this report were found at the same locality, J.M. 90, a site in the basal deposits of the Chemeron Beds which lie to the west of Lake Baringo in the Republic of Kenya. The beds have been described (see McCall *et al.*, 1967) and appear to be of lower Pleistocene age. Specimens of a primitive *Elephas*,

Anancus and *Deinotherium* have been collected from the same site, together with various other extinct genera.

Both specimens are remarkably complete; a skull and mandible of a *Papio*, damaged in the basi-occipital region and the skull and mandible of a colobid, also damaged in the basi-occipital region. An almost complete skeleton was found in articulation with the colobid skull while the *Papio* specimen was found without any associated post-cranial material. Dr. W. W. Bishop was responsible for the discovery of the colobid skeleton.

A mould was made of the colobid specimen *in situ* and casts have since been prepared, one of which is displayed in the National Museum, Nairobi, Kenya. The original specimens are stored at the National Centre for Prehistory and Palaeontology, Nairobi.

DESCRIPTION OF THE NEW COLOBID SPECIMEN

Order PRIMATES
Superfamily **CERCOPITHECOIDEA**
Family *COLOBIDAE*
Genus and Species *Paracolobus chemeroni* gen. et. sp.nov.

Diagnosis

A cercopithecoid of large size. The skull and postcranial material exhibit characteristics which are closer to *Colobus* than to *Papio*, *Simopithecus* and other large, Lower Pleistocene Cercopithecoidea.

The skull is broad muzzled and of medium length. The post-orbital constriction is marked, with weak temporal lines forming the lateral margin of the constricted area. The temporal lines unite at the bregma and there is no evidence of a prominently developed sagittal crest. The area posterior to the glabella between the temporal lines is markedly concave.

The dental arcade is rectangular with a long molar/premolar series. The upper incisors are small and there is a marked diastema between the upper lateral incisors and the canines. The premolars are large and have tall angular cusps with deep valleys. The molars have tall angular cusps with deep valleys between and the lower third molar bears a strong talonid.

The ascending rami of the mandible are high and set vertical to the long axis of the mandibular corpus. The posterior portion of the rami bears an inverted triangular depression from which one can assume that the post-glenoid process was long and robust. The mandibular corpus is deepest in the region of the molars.

The scapula has a distinctly concave costal aspect with a marked keel forming the lateral margin. The deltoid tuberosity of the humerus is poorly developed and the trochlea is shallow. The olecranon process of the ulna is inclined anterior to a line drawn along the axis of the shaft. There is a markedly concave area between the medial crest of the olecranon and the medial border of the trochlear notch.

Material

The basi-occipital and occipital regions of the skull have not been preserved. The mandible is complete, although slightly distorted by a crack which passes through the region of the symphysis.

One foot, both hands and the distal portions of the radii and ulnae are lacking from the post-cranial material which was found in articulation and connected to the skull.

Although the specimen is that of an adult male, the dentition is not worn down and details of cusps are well preserved.

Description of the skull

The muzzle is broad and of medium length. The supra-orbital tori curve upwards and backwards from the glabella. The temporal lines unite just posterior to the bregma and become very weak posterior to the marked post-orbital constriction. The area posterior to the glabella and between the temporal lines above the post-orbital constriction is markedly concave. The maxillary fossae are very poorly developed and could well be considered absent. The supra-orbital notch is weak. There is a marked diastema between the upper canines and the lateral incisors. The maxillary ridges are weak and converge towards the nasal bones which are very short. The nasal aperture is long and narrow. The nasal process of the premaxilla is remarkably long, extending backwards to the posterior margin of the nasal aperture. The zygomatic process begins above the first molars and the malar-maxillary area below the orbits is heavily constructed and wide antero-posteriorly. The muzzle has a rounded coronal section resulting from the inflation of the lateral aspects of the maxilla.

Description of the mandible

The posterior and anterior edges of the ascending rami are vertical to the corpus of the mandible. The vertical height of the condyles is about 64% of the overall length of the mandible.

The articular surface of the mandibular condyle does not extend on to the posterior aspect of the condyles. The surface faces upwards and the condyle is set so that the median aspects are inclined slightly backwards. The articular surface is narrow antero-posteriorly. The posterior surface of the ascending ramus bears an inverted triangular depression in the region just below the lip of the condyle. The mandibular corpus is deeper in the region of the third molars than in the region of the premolars.

The gonion region is not appreciably thickened but there is a prominent, inward-facing tuberculum on the inner side of the mandibular angle.

5

The region of the symphysis between the premolars is relatively deep. The mandibular fossae are very poorly developed and the channel for the buccinator muscle posterior to the third molars is marked and extends on to the inner surface of the rami.

Dentition

The molar and premolar teeth have tall, angular cusps with deep valleys. There is a weakly-developed longitudinal crest between the lingual cusps of the upper molars and the buccal cusps of the lower molars. The lower third molar has a well-developed talonid which is about half the length of the main part of the tooth. The premolars are large and fairly high-crowned. The upper incisors are small, but larger than the lower incisors; both are fairly high-crowned. The upper and lower canines are not particularly high-crowned nor very massive.

TABLE I

Measurement on the skull

Height of muzzle, posterior to M^1 at the alveolar margin	40 mm
Length of muzzle; nasion to alveolar point	70 mm
Breadth between lateral surfaces of maxilla at the alveolar margin above the third molars	55 mm
Breadth at M^1	53 mm
Breadth at Canines	54·5 mm
Minimum inter-orbital width	21 mm
Minimum width at post-orbital constriction	57·5 mm
Length of nasal aperture	39 mm
Maximum width of nasal aperture	10 mm
Length of nasal bones	22·5 mm
Maximum height of orbit	24 mm
Maximum width of orbit	20 mm
Diastema	5 mm
Length of molar-premolar series	52·5 mm
Width of upper incisor series	27·5 mm

TABLE II

Measurements on the mandible

Vertical height of condyle above the lower margin of the mandibular corpus .	79·5 mm
Maximum length of mandible—intra-dental point to the posterior edge of ascending ramus	125·5 mm
Maximum length of symphysis in median plane	46·5 mm
Maximum thickness of mandibular corpus below the third molar . .	13·5 mm
Length of molar-premolar series	61·5 mm

TABLE III

Measurements on the dentition

	UPPER				LOWER	
	Length	Breadth			Length	Breadth
I^1	6·5 mm	5·5 mm	I_1		5·0 mm	6·25 mm
I^2	6·0 mm	6·0 mm	I_2		5·0 mm	6·25 mm
C	13·6 mm	11·5 mm	C		11·0 mm	10·5 mm
P^3	8·0 mm	8·5 mm	P_3		12·0 mm	8·0 mm
P^4	8·0 mm	9·5 mm	P_4		9·5 mm	8·0 mm
M^1	10·5 mm	10·0 mm	M_1		11·0 mm	8·25 mm
M^2	12·0 mm	11·5 mm	M_2		12·5 mm	10·0 mm
M^3	13·5 mm	11·5 mm	M_3		16·5 mm	10·0 mm

NB. The teeth are set in such close proximity to one another that it has been impossible to obtain precise measurements on their length. The above table, therefore, should only be taken as an approximation of the actual dimensions.

Post-cranial material

In this report it has not been possible to present a detailed decription of all the material together with a comprehensive comparative analysis of the post-cranial material of other genera and species. The major morphological characteristics of the larger bones have been noted and, where possible, comparisons have been drawn to demonstrate some of the main diagnostic features of the material. Where comparisons have been made with living genera, a series of 12 of each have been used in an attempt to determine the individual variability of certain characteristics. The morphological features that have been demonstrated in this paper appear to be diagnostic and not to vary with individuals to any major extent.

Scapulae

The right scapula is more complete than the left, but both have been damaged. The costal surface is concave with a distinct longitudinal hump running down the centre. The lateral border is separated from the costal surface by a distinct keel. The concavity of the costal surface is particularly marked in the region proximal to the glenoid process.

The lateral aspect of the scapula is wide as a result of the keel and is at right angles to the plane of the dorsal and costal surfaces. A sharp ridge runs from the region of the neck of the glenoid process along about two-thirds of the overall length of the lateral margin.

The spinoglenoid notch is marked but open. The terminal portion of the acromion

TABLE IV

	Paracolobus chemeroni	Simopithecus oswaldi	Colobus polykomos	Papio cynocephalus	Cercopithecoides williamsi
1.	Fairly short muzzle.	Medium length muzzle.	Very short muzzle.	Long muzzle.	Very short muzzle.
2.	Weak temporal crests united just posterior to bregma. No evidence of sagittal crest.	Very marked temporal crests uniting anterior to, or at, the bregma. Marked sagittal crest.	Developed temporal crests which do not meet but pass posteriorly to join the nuchal crest.	Weak temporal crests uniting posterior to bregma. Poorly-developed sagittal crest.	Marked temporal crests uniting posterior to bregma or continuing separately to the nuchal crest. No evidence of a sagittal crest.
3.	Poorly-developed supra-orbital notch.	Well-developed supra-orbital notch.	Poorly-developed supra-orbital notch.	Well-developed supra-orbital notch.	Well-developed supra-orbital notch.
4.	Marked diastema between upper canines and lateral incisors.	Usually no diastema; if present, very slight.	Marked diastema between upper canines and lateral incisors.	Marked diastema.	Diastema between upper and lateral incisors and canines.
5.	Posterior and anterior edges of ascending ramus vertical to corpus of mandible. Ascending ramus high.	Posterior edge of ascending ramus almost vertical to corpus. Ascending ramus high.	Posterior and anterior edge of ascending ramus vertical to the corpus. Ascending ramus fairly high.	Posterior and anterior edges of ascending ramus set at an angle away from mandibular corpus. Ascending ramus low.	Ascending ramus inclined at an angle from the long axis of the mandibular corpus. Ascending ramus of medium height.
6.	Channel for buccinator muscle marked and passing on to the inner surface of the ascending ramus.	Channel for buccinator muscle very marked and passing on to the inner surface of the ascending ramus.	Poorly-developed channel for buccinator muscle. Does not pass on to inner surface of the ascending ramus.	Poorly-developed channel for buccinator muscle. Does not pass on to the inner surface of the ascending ramus.	Channel for buccinator muscle marked and passing on to the inner aspect of the ascending ramus.
7.	Mandibular and maxillary fossae very poorly developed.	Mandibular and maxillary fossae poorly developed.	No mandibular or maxillary fossae.	Strongly-developed maxillary and mandibular fossae in most adult specimens.	Maxillary and mandibular fossae very weak or absent.

	A	B	C	D	E
8.	Triangular depression below the posterior edge of the mandibular condyle.	Triangular depression below the posterior edge of the mandibular condyle.	Triangular depression below the posterior edge of the mandibular condyle.	No depression below the posterior edge of the mandibular condyle.	No depression below posterior edge of the mandibular condyle.
9.	Fairly small, high-crowned incisors.	Small, low-crowned incisors.	Fairly small, high-crowned incisors.	Large, high-crowned incisors.	Small, low-crowned incisors.
10.	Malar process begins just posterior to P^4.	Malar process begins posterior to M^1.	Malar process begins just posterior to M^1.	Malar process begins posterior to M^2.	Malar process begins above posterior aspect of M^1.
11.	Pronounced upward and backward curve of supra-orbital tori.	Slight backward curve to supra-orbital tori.	Slight downward and backward curve to supra-orbital tori.	Slight downward and backward curve to supra-orbital tori.	Slight upward, but pronounced backward, curve to supra-orbital tori.
12.	Markedly short nasal bones.	Fairly short nasal bones.	Short nasal bones.	Long nasal bones.	Short nasal bones.
13.	Lightly constructed mandible; corpus deeper in region of molars than in region of premolars.	Heavy mandible; corpus deeper in region of M_3 than in region of premolars.	Light mandible; corpus of mandible deeper in anterior region than in region posterior to M_2.	Light mandible; corpus deeper in anterior region than in region of M_2.	Relatively robust mandible; not markedly deeper in either anterior or posterior regions of the corpus.
14.	Poorly-developed longitudinal crests between lingual cusps of upper molars and buccal cusps of lower molars.	Pronounced longitudinal crests between lingual cusps of upper molars and buccal cusps of lower molars.	No longitudinal crests between cusps.	No longitudinal crests between cusps.	Slight longitudinal crests between buccal cusps of lower molars and lingual cusps of upper molars.
15.	Upper and lower premolars relatively large.	Upper and lower premolars relatively small.	Upper and lower premolars relatively large.	Small upper and lower premolars.	Proportionately large upper and lower premolars.

has been damaged so that it is not possible to ascertain its form. The part of the acromion that is preserved curves downwards quite distinctly.

The spine of the scapula begins as a pronounced ridge on the dorso-lateral aspect of the acromion and terminates at the medial border. The curvature of the spine is median, a character which is seen in the *Colobus* specimens examined but is absent in *Papio*.

The supraspinous region is distinctly smaller in area than the infraspinous region. The surface is concave in the region of the superior angle.

The coracoid process, the acromion and glenoid process do not appear to bear any diagnostic characteristics which would be useful for the purpose of this report.

TABLE V

Comparison of the scapulae of *Paracolobus chemeroni*; *Colobus polykomos* and *Papio cynocephalus*

Paracolobus chemeroni	*Colobus polykomos*	*Papio cynocephalus*
Costal surface concave with longitudinal ridge.	Costal surface concave with longitudinal ridge.	Costal surface generally flat.
Lateral border sharp and separated from costal surface by distinct keel which forms broad lateral aspect at right angles to the dorsal and costal surfaces.	Lateral border sharp and separated from costal surface by distinct keel which forms broad lateral aspect at right angles to the dorsal and costal surfaces.	Lateral border sharp, meeting costal surface over a rounded ridge. No keel forming lateral margin to costal surface.
Spine of scapula terminates at medial border of scapula.	Spine of scapular terminates at medial border of scapula.	Spine of scapula terminates before medial border.

Humeri

Both humeri are complete and well preserved. The bicipital groove is poorly developed and its anterior lip is a rounded ridge which continues down the shaft as the anterior border of the bone to the region of the deltoid tuberosity. The deltoid tuberosity is very poorly developed and totally unlike that seen in *Papio*, *Colobus* or *Simopithecus*.

The lateral aspect of the shaft in the proximal half of the bone is rounded with only a slight suggestion of a ridge in the region below the deltoid tuberosity.

The medial aspect of the shaft in the proximal half of the bone is similarly rounded with no suggestion of the ridge-like medial border that is seen in *Colobus*.

The distal portion of the shaft is compressed antero-posteriorly and the lateral border in this region is sharp and ridge-like. The medial border is rounded.

The olecranon fossa is not deep and the surface of the lateral epicondyle meets the articular surface of the trochlea as a slight ridge. The medial epicondyle projects well beyond the medial border of the trochlea; the epicondylar groove is not well developed.

TABLE VI

Overall length of humerus measured with the posterior aspect resting on horizontal plane . . . . 256 mm
Maximum width of distal end across epicondyles and at right angles to the long axis of the bone . . . . 45 mm
Bilateral width at proximal end . . . . 34 mm
Minimum antero-posterior diameter of shaft . . . . . . 17·5 mm

TABLE VII

Comparison between the humeri of *Paracolobus chemeroni, Colobus polykomos* and *Papio cynocephalus*

Paracolobus chemeroni	*Colobus polykomos*	*Papio cynocephalus*
Bicipital groove poorly developed.	Bicipital groove marked.	Bicipital groove marked.
Deltoid tuberosity weak.	Deltoid tuberosity marked.	Deltoid tuberosity marked.
Medial and lateral borders of proximal end of shaft are rounded.	Medial and lateral borders of proximal end of shaft sharp and ridge-like.	Medial border ridge-like, lateral border rounded in the proximal region of shaft.
The medial epicondyle projects well beyond the medial border of the trochlea; the epicondylar groove is not well developed.	The medial epicondyle projects well beyond the medial border of the trochlea; the epicondylar groove is not well developed.	Medial epicondyle is not extended.
On posterior aspect, the articular surface of the trochlea is shallow and only a slight ridge is formed by the junction with the surface of the medial epicondyle.	On posterior aspect, the articular surface of the trochlea is shallow and only a slight ridge is formed by the junction with the surface of the medial epicondyle.	The posterior aspect of the articular surface of the trochlea is deeply set between epicondyles. High and marked ridge formed at junction of articular surface and the medial epicondyle.

Ulnae

Both ulnae were collected but unfortunately the distal portions were not present. The olecranon process is robust and, apart from being medially inclined to a certain extent, it lies anterior to the long axis of the shaft. A distinct ridge forms the medial border of the olecranon and between this ridge and the medial limit of the articular surface of the trochlear notch there is a marked depression. The radial notch is deep and has a continuous articular surface. Below the radial notch there is a very marked fossa.

The supinator crest is marked and there is a shallow groove between it and the posterior border of the shaft for a distance of about 80 mm. The anterior border of the shaft in the region immediately below the coronoid process is sharp.

TABLE VIII

Comparison between the ulnae of *Paracolobus chemeroni*; *Colobus polykomos* and *Papio cynocephalus*

Paracolobus chemeroni	*Colobus polykomos*	*Papio cynocephalus*
Olecranon process inclined slightly anteriorly in relation to long axis of shaft.	Olecranon process inclined anteriorly in relation to long axis of shaft.	Olecranon process inclined posteriorly in relation to long axis of shaft.
Marked depression between medial crest of olecranon and medial limit of trochlear notch.	Marked depression between medial crest of olecranon and medial limit of trochlear notch.	No medial crest on olecranon and no depression between olecranon and medial aspect of trochlear notch.
Marked radial notch with single articular surface.	Marked radial notch with single articular surface.	Marked radial notch. Articular surface usually in two parts.

Radii

Both radii were collected but the distal ends had been broken off prior to excavation. The radius is distinctly robust. The tuberosity of the radius is very well developed. A little below the tuberosity, the medial surface of the shaft becomes remarkably flat, the anterior border being sharp initially and becoming rounded towards the middle and distal portions of the shaft. The posterior border of the shaft is rounded.

TABLE IX

Comparison of the radii of *Paracolobus chemeroni, Colobus polykomos* and *Papio cynocephalus*

Paracolobus chemeroni	Colobus polykomos	Papio cynocephalus
Pronounced tuberosity of radius.	Pronounced tuberosity of radius.	Tuberosity less pronounced.
Medial surface flat.	Medial surface generally flat with slight groove in mid-portion of shaft.	Medial surface bears marked groove between anterior and posterior borders.
Rounded borders to shaft in the distal region.	Borders angular.	Borders angular.
Radius markedly robust.	Radius slender.	Radius generally slender.

Femora

Both femora were recovered complete and in very good condition. Morphologically, there is very little to distinguish the femora of *Paracolobus chemeroni* from those of other genera. There are a few minor points, particularly in the region of the proximal end, but these characters appear to vary between individuals and it would be unwise to stress them in the absence of additional specimens.

TABLE X

Measurements on the femur

Maximum length—anterior aspect resting on a horizontal surface . . .	272 mm
Bicondylar width at right angles to long axis of shaft	48·5 mm
Diameter of shaft at mid-point in antero-posterior direction . . .	21 mm

Tibiae and fibulae

Both tibiae and fibulae were recovered. As in the case of the femora, the few morphological characteristics that appear to be distinctive are so slight that it would be unwise to place stress on them for purposes of comparison with other genera. The fibulae are slightly crushed and it is not possible to obtain accurate measurements at the present time.

Table XI

Measurements on the tibia

Maximum length of tibia	247 mm
Bilateral width of shaft in region of tuberosity	18 mm
Antero-posterior breadth at tuberosity	30 mm
Bilateral width across condyles	43 mm

Other post-cranial material

The vertebral column and both innonimates were collected but they are not included in this report as the material has still to be fully developed from its matrix. The pes and a few other small bones are not dealt with in the present report.

Discussion

The fossil Cercopithecoidea of both East and South Africa are generally represented by cranial remains and for this reason the skull and mandible of *Paracolobus chemeroni* are of particular interest. It should be noted, however, that the morphology of the post-cranial material is of great importance and can assist greatly in an accurate classification of the group being studied.

For comparative purposes, a series of 12 specimens of *Colobus polykomos* and *Papio cynocephalus*, the available East African *Simopithecus* material and the Cercopithecoidea material from some of the richer Lower Pleistocene deposits in South Africa have been carefully examined. At the time of writing this report there are no East African fossils of comparable morphology available.

In Table IV some of the most diagnostic morphological characteristics have been compared to the corresponding characteristics of *Colobus polykomos*, *Simopithecus oswaldi*, *Papio cynocephalus* and the South African *Cercopithecoides williamsi*.

In each case the characters noted appear to be common to that genus and may be taken as diagnostic, especially when used as a series of characteristics. Sexual dimorphism is known to be an important factor and the features used in the comparative tables of this report are those which appear to vary least between individuals provided that the specimens are of a similar age. It must be stressed, however, that individual characteristics are of little value when used on their own and that failure to appreciate this may well have led to some of the taxonomic errors that have plagued the classification of fossil Cercopithecoidea.

Tables V, VII and IX have been prepared in an attempt to demonstrate some of the major morphological characteristics of certain limb bones that were found in articulation with the cranium of *Paracolobus chemeroni*. Fossil material available for purposes of comparison is fragmentary and scarce and, consequently, *Colobus polykomos* and *Papio cynocephalus* have been used in the comparative tables.

As in the case of the cranium, a good series of specimens have been examined in order to determine the significance of various characteristics. Those that are used are considered to represent some of the major diagnostic features which do not appear to vary greatly between different sexes of a similar age group within a genus.

Papio cynocephalus has been used to demonstrate the morphology of the larger terrestrial Cercopithecoidea while *Colobus polykomos* demonstrates the anatomical characteristics of the highly specialized arboreal Colobinae. The fossil post-cranial remains from the various Lower Pleistocene sites in both eastern and southern Africa are limited and often fragmentary and appear to represent the terrestrial Cercopithecoidea such as *Simopithecus, Parapapio* and *Papio*. In the course of the preparation of this report the major portion of the specimens from the cave breccias of southern Africa were examined and no examples of the post-cranial material of any arboreal Cercopithecoidea came to light.

The skull and mandible of *Paracolobus chemeroni* show certain characteristics which are undoubtedly attributable to the sub-family Colobinae. With the added evidence of the morphology of the post-cranial material there is little doubt as to the validity of placing the specimen in the Colobinae.

The *Cercopithecoides williamsi* material from southern Africa is of particular interest. The dental characteristics are very similar to those seen in *Paracolobus chemeroni*, although the latter is considerably larger in every proportion. A few of the cranial characteristics are common to both genera but, on the whole, *Cercopithecoides williamsi* is much closer to *Colobus* than to *Paracolobus*. It is regrettable that at present there is no post-cranial material since this would greatly assist in the final analysis of the *Cercopithecoides williamsi* specimens from southern Africa.

Conclusion

The morphological evidence provided by the skull, mandible and post-cranial material is quite distinctive and shows that *Paracolobus chemeroni* is different from any of the other known fossil or living Cercopithecoidea. Based on this evidence the specimen has been placed within the Colobinae and given the name of *Paracolobus chemeroni*.

DESCRIPTION OF A NEW SPECIES OF PAPIO

Order PRIMATES
Superfamily **CERCOPITHECOIDEA**
Family ***CERCOPITHECIDAE***
Genus and species ***Papio baringensis*** sp. nov.

Diagnosis

A true *Papio* of large size. The muzzle is long and narrow, the maxillary ridges diverging slightly posteriorly to join the lower lateral margin of the orbits. Marked infra-orbital foramina. The post-orbital constriction is very marked, the temporal crests are strong and unite anterior to the bregma.

The mandible is long and the posterior border of the ascending ramus in the region below the condyle is rounded. There is a pronounced mandibular fossa. The anterior aspect of the symphysis slopes almost vertically.

The canines are high-crowned while the incisors are small and low-crowned. A pronounced groove is present on the mesial aspect of the canines.

Material

The skull and mandible are of a fully adult male individual. The preservation of the bone is excellent with minimal distortion. The occipital and basi-occipital regions have been lost due to a fracture through the parietals down to the posterior margin of the zygomatic arches. The two sides of the muzzle are not bilaterally symmetrical owing to a crack traversing the right orbit down to the maxilla and palate.

The mandible has suffered slight damage along the inferior margin of the right ascending ramus. The left canine is broken off, a little above the alveolar margin. A slight bilateral distortion has been caused by a crack through the right mandibular corpus at a point just posterior to the canine.

Description of the skull

The muzzle is long. The maxillary fossae are very poorly developed with the lateral aspect of the maxilla being almost flat and meeting the malar process in a very gradual curve. The maxillary ridges are strongly developed and diverge posteriorly towards the infra-lateral margin of the orbits. The infra-orbital foramina are strongly developed. The post-orbital constriction is very marked indeed, the minimum width of the muzzle being greater than the width of the post-orbital constriction.

The supra-orbital ridges are pronounced and the supra-orbital notch is well developed with a suggestion of spines having been present. The temporal crests are strong and unite at a point about 20 mm anterior to the bregma. The crest continues posteriorly, suggesting that the specimen had a well-developed sagittal crest. The orbits are of greater height than width. The malar-maxillary area below the orbits is massive and wide antero-posteriorly.

There is a marked diastema between the lateral incisors and the upper canines. The canines have a pronounced groove on their mesial aspect. The incisors are very small and low-crowned.

The incisive fossa is small, being greater in length than in width. The post-glenoid process is robust but short and the articular tubercle is well developed.

Description of the mandible

The mandibular fossae are pronounced and the form of the mandible is typical of *Papio* in that the mandibular corpus is deeper in the region of the premolars than in the region posterior to the third molar. The symphysial region between the premolars is deeply excavated. The ascending rami are low and inclined at an angle away from the long axis of the mandible. The region of the gonion is not appreciably thickened.

The articular surface of the mandibular condyles is large and extends posteriorly downwards to give a surface that faces both upwards and backwards. The condyles are large antero-posteriorly as well as being wide laterally. The channel for the buccinator muscle is well developed but does not extend on to the inner surface of the ramus. The lower anterior aspect of the symphysis is almost vertical below the incisors and is distinct from the living *Papio cynocephalus* where the lower anterior aspect of the symphysis slopes backwards. On the antero-lateral aspect of the horizontal ramus below the canines there is a marked roughened ridge running in an antero-posterior direction.

TABLE XII

Measurements on the skull

Height of muzzle from the alveolar margin posterior to the upper third molar	58·5 mm
Length of muzzle; alveolar point to nasion	112 mm
Breadth of maxilla taken at alveolar margin at the third molars	53 mm
Breadth of muzzle at canines	54·5 mm
Minimum inter-orbital width	12·5 mm
Length of the nasal bone	60 mm
Maximum height of orbit	27·5 mm
Maximum width of orbit	24 mm
Minimum width at post-orbital constriction	43·5 mm
Diastema	7 mm
Length of molar-premolar series	50 mm
Width of incisor series	28·5 mm

TABLE XIII

Measurements on the mandible

Length of molar-premolar series	64 mm
Length of mandible: intra-dental point to the posterior margin of the ramus	159 mm
Vertical height of condyle with mandibular corpus resting on a horizontal surface	77·5 mm
Length of symphysis in the median plane	55 mm
Maximum thickness of corpus below M$_3$	16·5 mm

TABLE XIV

Measurements on the dentition

	UPPER			LOWER	
	Length	Breadth		Length	Breadth
I^1	7·0 mm	6·0 mm	I_1	6·0 mm	5·0 mm
I^2	6·0 mm	6·0 mm	I_2	6·5 mm	5·25 mm
C	16·5 mm	13·0 mm	C	15·0 mm	8·5 mm
P^3	7·0 mm	8·0 mm	P_3	24·0 mm	7·0 mm
P^4	8·5 mm	8·5 mm	P_4	9·0 mm	7·0 mm
M^1	10·5 mm	10·0 mm	M_1	11·5 mm	8·5 mm
M^2	13·0 mm	12·0 mm	M_2	12·5 mm	10·0 mm
M^3	13·5 mm	12·0 mm	M_3	17·0 mm	11·0 mm

Discussion

The new fossil from the Chemeron Beds is without doubt a *Papio*. Very little information is available at present on other material from the Lower Pleistocene sites in East Africa but the reverse is true in southern Africa where a great many specimens have been collected and described.

During the preparation of this report many of the fossil *Papio* from southern Africa were examined and it was concluded that the specimen from Kenya is definitely distinct morphologically. Perhaps the most striking points of difference are the extreme post-orbital constriction and the length and width of the muzzle.

A maxillary fossa is absent although the mandibular fossa is very pronounced. This feature is known to vary considerably in the living *Papio cynocephalus* and for this reason the point has not been stressed as an important morphological characteristic.

Conclusions

The specimen described in this report has various anatomical characteristics which are distinct from the known fossil *Papio* from both eastern and southern Africa. Based on a series of comparisons with both fossil and modern *Papio* material, the specimen has been allocated the specific rank *Papio baringensis*.

REFERENCES

FREEDMAN, L. 1957. The Fossil Cercopithecoidea of South Africa. *Ann. Transv. Mus.* **23**: Part II, 121.

—— 1960. Some New Cercopithecoid Specimens from Makapansgat, South Africa. *Palaeont. afr.* **7**.

LEAKEY, L. S. B. 1943. Notes on *Simopithecus oswaldi* Andrews from the Type Site. *Jl. E. African nat. Hist. Soc.* **17**: 39.

LEAKEY, L. S. B. and WHITWORTH, T. 1958. Notes on the Genus *Simopithecus*, with a Description of a New Species from Olduvai. *Coryndon Memorial Museum Occasional Papers*, No. 6.

McCALL, G. J. H., BAKER, B. M. and WALSH, J. 1967. Late Tertiary and Quaternary Sediments of the Kenya Rift Valley. In *Background to Human Evolution in Africa*. W. W. BISHOP and J. D. CLARK, Ed. University of Chicago Press.

LEGENDS TO PLATES

PLATE 1
Paracolobus chemeroni ♂

PLATE 2
Papio baringensis ♂

PLATE 3
Cercopithecoides williamsi ♂

PLATE 4
Colobus polykomos ♂

PLATE 5
Papio cynocephalus ♂

PLATE I

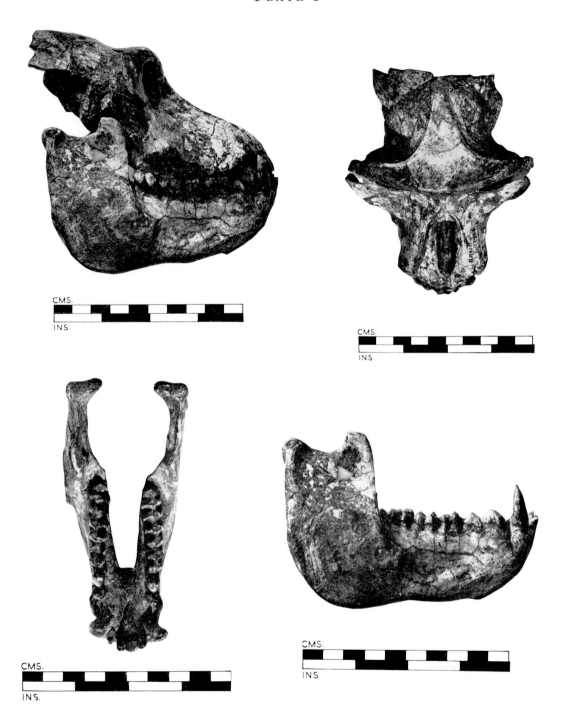

PLATE 2

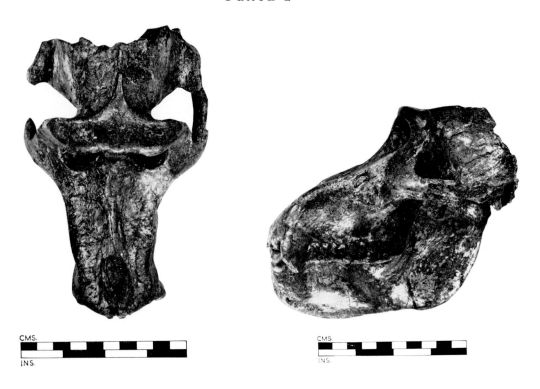

CMS.

INS.

CMS.

INS.

CMS.

INS.

PLATE 3

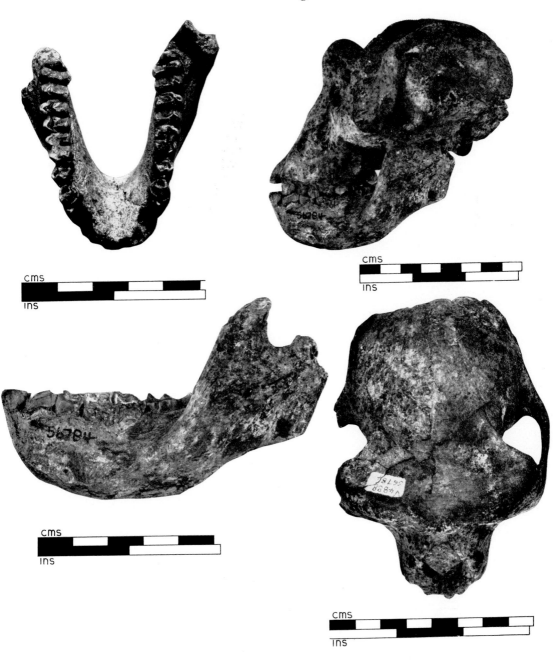

cms

ins

cms

ins

cms

ins

cms

ins

PLATE 4

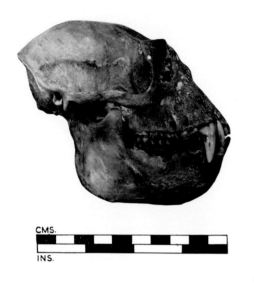

CMS.

INS.

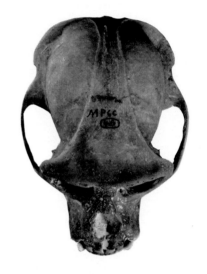

CMS.

INS.

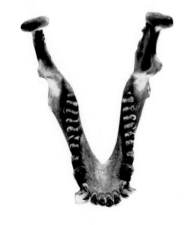

CMS.

INS.

CMS.

INS.

PLATE 5

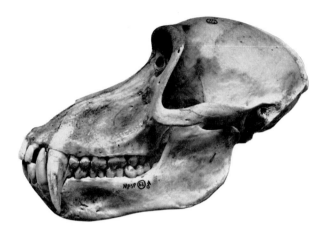

CMS.

INS.

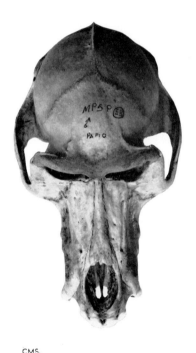

CMS.

INS.

CMS.

INS.

PLEISTOCENE EAST AFRICAN RHINOCEROSES

D. A. HOOIJER

Rijksmuseum van Natuurlijke Historie, Leiden, Holland

INTRODUCTION

Though a good few names have been given, there are virtually only two species of rhinoceroses in the Quaternary of East and South Africa. These are the black rhinoceros *Diceros bicornis* (L.) and the white rhinoceros *Ceratotherium simum* (Burchell). The former a browser, the latter a grazer, the skulls appear markedly distinct; yet from the specializations among the Pontian (Early Pliocene) forms of *Diceros* it seems likely that the emergence of *Ceratotherium* from the *Diceros* stock took place only in the Pliocene (Thenius, 1955). By Early Pleistocene times the two living forms had already diversified, and their teeth are nearly as distinct as those in modern skulls. The distinction is most easily made in the permanent upper dentition, and the milk dentition shows progressive divergence in pattern as we pass backward along the series (Hooijer, 1959). *Ceratotherium simum* is the hypsodont form, with much cement on the teeth, and with the transverse lophs in the upper

6

molars (protoloph and metaloph) obliquely placed, recurved backward. This does not show in *Diceros bicornis*, which also has less high-crowned teeth.

In the Early Pleistocene form of *Ceratotherium simum* protoloph and metaloph are less oblique, the metaloph in fact still transverse in its course, and there is evidence that the crowns were not quite as high as those at present either. This has already been demonstrated by various authors. As we shall see in the present paper *Diceros bicornis* also displays a trend toward progressive hypsodonty in the course of the Quaternary. The evidence is scanty, resting upon a few isolated observations, and we could do with a great deal more material, but the trends are evident. There is subspecific advance at least in the teeth of both of the extant African species of rhinoceros. The history of them is one sad story of slaughter, intensified by the demand for their horns which are highly valued for their supposedly aphrodisiac properties in the Far East, where the Asiatic species of rhinoceros are so rapidly becoming exterminated that horns are much harder to come by.

Diceros bicornis, the least threatened species with a number of individuals of well over 10,000, shows some geographic variation, which has recently been reported by Groves (1967, p. 270) thus: " A single widespread race is indicated, distributed from Transkei in the south to Lake Victoria in the north, with a very large race at the Cape, another large one on the Chobe river, two smaller ones respectively in East Africa to the east of the Rift Valley and in West Africa, and a large broad-skulled one to the northwest of the Kenya distribution." Such a pattern of geographic subspeciation must have existed at any time in the Pleistocene as well, but most of the meagre material that we have in the way of fossils are teeth, in which these racial differences do not show.

Ceratotherium simum, not even 2,000 individuals of which survive today, has a discontinuous historic distribution area: south of the Zambesi, and then again in Equatorial Africa west of the Nile (maps in Heller, 1913, pl. 10, and Guggisberg, 1966, p. 79). There is remarkably little difference between these recent geographical forms: this concerns mainly the greater depth of the dorsal concavity of the adult skull in the South African specimens (at least 60 mm as compared with 50 mm or less in the Lado Enclave specimens), and further the greater length of the permanent toothrow in the South African specimens, averaging about 300 mm as against 270 mm in the northern skulls (Heller, 1913, p. 30). An adult Leiden Museum skull from the Cape (cat. ost. b) bridges the gap, with a depth of dorsal skull concavity of 55 mm. The difference in average length P^2–M^3 is trivial, and no indication of general size, as the overall length of the toothrow decreases during life as a result of interproximal wear. A subadult skull from the Cape in the Leiden Museum (cat. ost. a, Van Horstok, August, 1831) has 250 mm for the length P^2–M^2 in a slightly worn state (M^3 erupting), whereas in cat. ost. b, in which the teeth are well worn (including M^3) the length P^2–M^2 has reduced to 210 mm, a difference of 40 mm. Such differential characters as have been advanced are hardly worthy of subspecific distinction, yet the two subspecific names, *C. simum simum* (Burchell) for the southern, and *C. simum cottoni* (Lydekker) for the northern form, are being universally upheld in the literature to this day.

The South African remains of Quaternary rhinoceroses known to date have been revised by Cooke (1950), and the Late Pleistocene remains of the black and the white rhinoceros from Hopefield, Cape Province, have been described by Hooijer and Singer (1960). In Late Pleistocene deposits the black rhinoceros is the common form, and the white rhinoceros rare. Contrariwise, in the Early Pleistocene Makapansgat Limeworks Cave collection (Hooijer, 1959) the white rhinoceros predominates. The collection from this australopithecine site is very remarkable in that it consists almost exclusively of milk molars, most of them unworn or hardly touched by wear. This is an indication that it was mainly sucklings that became fossilized in this cave; such a concentration of milk teeth has not before been recorded from any rhino site, in Africa or elsewhere. To me it seems evidence that *Australopithecus* was capable of a high level of cooperative behaviour for hunting; this only corroborates the views on the cultural status of the australopithecines currently based on other evidence. No differences between the Pleistocene and the Recent teeth were found apart from a tendency for the fossil teeth to be larger than the corresponding modern, which is a common phenomenon (Hooijer, 1950).

Turning now to the East African Pleistocene record, both the black and the white rhinoceros have been found at Olduvai Gorge, Tanzania. *Diceros bicornis* does not occur in the lower portion, the Early Pleistocene Bed I and Lower Bed II, but is found only higher up. It does occur in the Early Pleistocene of Makapansgat in South Africa, though sparingly, and, as we shall see in the sequel, we now have evidence of its presence in the Early Pleistocene deposits of the Omo Basin in Ethiopia as well. This early form of *D. bicornis* differs from the extant form only in being slightly larger-toothed, and less hypsodont. We know nothing of its skull characters yet. The modern genus *Diceros* appears first in the Early Pliocene, in North Africa as well as in Europe. I have recently described a small, extinct rhinoceros from the Late Miocene of Fort Ternan, Kenya, that appears to be a collateral development of a browsing, bicorn rhinoceros, differing from *Diceros* in a combination of primitive and progressive features: *Paradiceros mukirii* Hooijer (1968). It is more advanced than the Early Miocene forms previously described (Hooijer, 1966).

The white rhinoceros ranged all over the African continent in the Pleistocene. The North African " *Rhinoceros mauritanicus* " of Pomel's, as Arambourg (1938, p. 22) has shown, is inseparable from *Ceratotherium simum*. The name *mauritanicum* should, however, remain available and used if and when a find of an entire skull of the Moroccan white rhinoceros would eventually prove it to be distinguishable from the living (sub)species. Arambourg (l.c.) stated it to be closer to the typical (southern) than to the Equatorial form in its great toothrow length (one observation: 300 mm), but this does not carry any weight as indicating any closer actual relationship of the Moroccan fossil with the South African rather than with the northern recent form, the size difference between which is trivial at any rate, as we have seen above. Moreover, Pleistocene remains of living mammalian species generally being on the large side, their measurements are bound to agree better with the larger than with the smaller of the observations on the living forms.

One of the new fossil forms described from Olduvai Gorge as a result of the German expedition of 1913 was *Rhinoceros simus germanoafricanus* Hilzheimer (1925), based on an incomplete skull and mandible. As has already been shown by Zeuner (1934, p. 63), the Olduvai white rhino is not more primitive in cranial features than the extant form. It is not known from which level at Olduvai the type specimen came; Leakey (1965, p. 25) writes that from the illustrations it seems that it may have come from deposits younger than Bed IV. However, the upper molars of *germanoafricanum* do differ from those of the living white rhinoceros: the metaloph is transverse in its course, and the protoloph at its origin is nearly perpendicular to the ectoloph and recurved backward only distally.

These peculiar characters clearly shown in Hilzheimer's illustration (reproduced in Arambourg, 1947, p. 299, fig. 24) are just the same as those found by Dietrich (1942, 1945) to be typical of the Early Pleistocene rhinoceros from Laetolil, Tanzania, which was baptized *Serengeticeros efficax* Dietrich. The resemblance in molar structure has been noticed by Arambourg (1947, p. 300) in his discussion of the Omo white rhinoceros, which also differs from the modern form in being less plagiolophodont, and he refers to the Omo form as *Atelodus* cf. *germanoafricanus*. Dietrich (1945, p. 56) based his diagnosis on the upper molars exclusively, the cranial characters being unknown, and did not fail to mention (Dietrich, 1945, p. 68) that the molars of Hilzheimer's skull could very well belong to *Serengeticeros*. A comparison between Dietrich's 1945 illustrations of the upper molars and Hilzheimer's figure indicates the virtual identity in molar patterns of the Olduvai and the Laetolil fossil rhinoceros.

These are not the only Early Pleistocene sites whence this primitive type of *Ceratotherium simum* molar comes: Hopwood (1926) described as *Rhinoceros scotti* an M^2 sin. from the Kaiso Beds of Uganda that is indistinguishable from *C. simum* but displays the primitive characters of the Laetolil and Omo molars (Hopwood, 1926, p. 17, fig. 3), as remarked by Dietrich (1945, p. 51) and by Arambourg (1947, p. 301).

Neither Arambourg (1947, p. 300) nor Leakey (1965, p. 25) deem it necessary to uphold the genus *Serengeticeros* of Dietrich's. Neither do I consider the difference between the (Early) Pleistocene and the living white rhinoceros to be of generic, or even specific value. It seems that subspecific distinction is the most that can be accredited to the Early Pleistocene stage of *C. simum*, which has been recognized, as we have seen, under three different names, at Kaiso, Laetolil, Olduvai, and Omo. The name that has priority is *germanoafricanum* Hilzheimer, 1925, antedating *scotti* Hopwood by one, and *efficax* Dietrich by seventeen years.

In his preliminary report on the Olduvai fauna, Leakey (1965, p. 25) writes that some Bed I and lower Bed II specimens seem to be *Ceratotherium efficax*, whereas a number of specimens from upper Bed II and Bed IV are placed under *Ceratotherium simum germanoafricanum*. I believe we cannot really differentiate the two, and that the name by which the primitive, Early Pleistocene stage should be known is *Ceratotherium simum germanoafricanum* (Hilzheimer), the name *efficax* Dietrich falling away as a synonym. Besides in being less plagiolophodont than the recent form, it is also slightly less hypsodont: Dietrich (1945, p. 59) mentions that the height of M^3 of "*Serengeticeros*" (reconstructed) is 85 mm as opposed to some 120–130 mm in the Late Quaternary *Ceratotherium*. Dietrich's reconstruction may be a little too low, although there remains a difference. An Omo M^3 is stated by Arambourg (1947, p. 297) to be close to 10 cm in height.

Rhinoceros specimens from Olduvai Gorge are present in great numbers at the National Museum Centre for Prehistory and Palaeontology in Nairobi, where I had the opportunity to study them in the summer of 1967 through the courtesy of Dr. L. S. B. Leakey. This study was aided by a grant from the Wenner-Gren Foundation for Anthropological Research, Inc., New York. N.Y. I have also included Olduvai material in the British Museum (Natural History) in London; Dr. A. J. Sutcliffe and Mrs. S. C. Coryndon, F. L. S., kindly made these specimens available. Surface material that cannot be allocated to any level in particular has been left out of account, unless the specimen is particularly well-preserved.

There is further rhinoceros material from the Early Pleistocene of Laetolil, Tanzania, at the Nairobi Centre as well as at the British Museum (Natural History). The rhinoceros specimens obtained in the Omo Basin, southern Ethiopia, by the Kenyan, French, and American parties working there in the summer of 1967 have also been studied by me; preliminary accounts of the results of the field work, with some faunal data, have recently been published (Arambourg *et al.*, 1967; Howell, 1968). Dr. W. W. Bishop has allowed me to study the rhino material from the Chemeron Formation, Lake Baringo, Kenya (see Martyn and Tobias, 1967). Finally, some material has been included originating from Kanam West, Naivasha, and Olorgesailie. I am grateful to all who have made this material available for study, and to Mrs. J. G. Ament who made the photographs.

CHEMERON FORMATION, LAKE BARINGO, KENYA

A skull of *Ceratotherium* found *in situ* in the Basal Beds of the Chemeron Formation, Lake Baringo (J.M.91) and a maxillary portion holding the much worn M^{1-3} sin. from the Lower Fish Beds of the same Formation (J.M.507) have been examined by me at the Nairobi Centre in 1967 through the courtesy of Dr. W. W. Bishop. The mammalian fauna as a whole, as Leakey (in Martyn and Tobias, 1967, p. 477) observed, is strongly suggestive of that from strata attributed to the Early Villafranchian. Both the skull and the palate indeed display the features of the primitive *Ceratotherium simum* described by Dietrich from Laetolil, here named *C. simum germanoafricanum* (Hilzheimer).

The skull (Pl. **1,** figs. 1–2) unfortunately is rather distorted, and lacks the front part; P³ dext. is the foremost tooth preserved. The palate is broken, and the left molar series pushed inward, M¹ mostly so, but the posterior border of the palate is on a level with that of M², as in the modern form. Only the right zygomatic arch and the right half of the occiput are preserved. The postglenoid processes are present, but the posttympanic processes broken off. The dorsal surface of the skull is missing. Very few measurements can be taken from a skull in this state of preservation (Table VIII); the bicondylar width is no larger than in recent skulls of *C. simum*, and the zygomatic width, which can be approximately taken, is about as large as in the largest recent skulls measured by Heller (1913: 373 mm and 384 mm, both old with M³ worn). Despite the deformations, the fossil skull appears to have the same backwardly inclined occiput by which the modern form is characterized, and cranial differences from recent *C. simum* are not apparent.

With the molars, however, it is a different matter. As can be seen most distinctly in the M¹ but also in the premolars and M² the metaloph is not oblique but rather transversely placed, while the backward recurvation of the protoloph is less marked than that in modern dentitions, i.e., the " *Serengeticeros* " characteristics discussed above. Moreover, the slightly worn M³ of the Chemeron skull appears to be less hypsodont than that of the modern form. Although the outer surface is exposed only for a height of 45 mm, it can be compared with other specimens and this comparison tends to show that the sides of the crown converge more markedly crownward in the Chemeron sample than in the others.

The outer surface of M³ of skull J.M.91 has an anteroposterior length at the alveolar border of 68 mm (this is not the actual base of the crown yet). The length at the occlusal surface, which is 45 mm higher up the crown, is only 38 mm, giving a decrease in outer surface length of 30 mm. Two recent M³ of *Ceratotherium simum* examined show that the crownward taper of the sides of the outer surface over a height of 45 mm is decidedly less, as follows. The M³ of a southern skull (Leiden Museum, cat. ost. b) has an exposed portion of just 45 mm in height, and the antero-posterior diameter of the outer surface is 73 mm basally, and 62 mm apically, giving a decrease of 11 mm. Another recent M³, in a skull from the Western Nile District, Uganda (Leiden Museum, reg. no. 13120), has an anteroposterior diameter of the outer surface at base of 68 mm, which has reduced to 58 mm 45 mm higher up, giving a decrease of 10 mm. Thus, although actual crown heights cannot be given, the fossil M³ is seen to be less hypsodont than the modern.

It is of interest to note that two Olduvai Bed II specimens of M³, which will be listed later, when measured in the same way, prove to be intermediate between the Chemeron M³ and the recent in degree of crownward taper, while a Bed IV Olduvai specimen conforms to the recent type. The Bed II specimens are M.14810, an M³ dext., and OLD/57, SHK II, 181, an M³ sin., with a decrease in outer surface length over 45 mm of height of 20–22 mm. The Bed IV specimen, OLD/55, BK II, 95, an M³ sin., tapers only 5 mm in anteroposterior outer surface length over a height of 45 mm. These few observations point to progressive heightening of the crown of M³ of *C. simum* from Early Pleistocene to Recent.

The maxillary portion from the Chemeron Formation (J.M.507) holds only the three molars, from the left side, and in M² the transverse position of the metaloph is clearly seen (Pl. **2**, fig. 1). While it is characteristic of *C. simum* molars for the crochet to unite with the crista so as to form a medifossette, exceptions do occur, and the present specimen is such an exception: the crochet extends across the medisinus without uniting with the crista. The molar is very much worn down, and the medisinus is cut off from the internal border of the crown. Of the M³ in the same specimen, worn down to a few mm from the crown base, only the central part and a small portion of the anterior face remain. The medisinus is likewise closed off internally, but the crochet extends completely across the valley, making for an elongated fossette. No further observations on this specimen can be given.

The Chemeron specimens, therefore, in their transverse position of the metaloph of the upper molars, agree with the Laetolil form, and can be identified as *C. simum germanoafricanum* just as the Olduvai, Omo, and Kaiso material discussed above. Although no cranial differences between the fossil and the living form have been observed, the degree of hypsodonty in the fossil form is less advanced than in the recent, as the observations on M³ have shown.

OLDUVAI GORGE, TANZANIA

Remains of rhinoceroses come from all levels at Olduvai. Most of it is teeth or parts of teeth of *Ceratotherium simum*; some specimens represent *Diceros bicornis*, and this has been marked as such in the lists that follow. The postcranial bones I have been unable to identify specifically; the two recent species are extraordinarily close osteologically although the white is the larger, which is of little help when dealing with Pleistocene bones. Tables of measurements of the postcranial skeletal remains are given at the end of the present paper; the recent skeletons of *C. simum* and *D. bicornis* compared are in the National Museum Centre for Prehistory and Palaeontology Department of Osteology, and their measurements have been given also in cases in which there were no corresponding bones in the fossil collection; they may be needed later.

The Olduvai material originates from the following sites (M. D. Leakey, 1965, and personal communication):

DK I—Lower Bed I; FLK NN—Mid Bed I; KK I—Bed I; MTK I—Bed I; MTK—Bed I or II; HWK(E)II—Lower Bed II; FLK II—Upper Bed II; EF-HR—Upper

Bed II; FC II—Upper Bed II; MNK II—Upper Bed II; SHK II—Upper Bed II; BK II—Upper Bed II; TK II—Upper Bed II; JK2, Geol. Pit 8—Bed III; JK2/A— Bed III or IV; JK2/B—Bed III or IV; MRC—Bed IV; Rhino Korongo—Bed IV; CMK—Bed IV.

The lower part of Bed II is separated from the upper part by a geological break; the lower part of Bed II belongs with Bed I to the Early Pleistocene (Late Villa-franchian), and the upper part of Bed II up to Bed IV represents the Middle Pleisto-cene (Leakey, 1965, p. 78). It has been found that some upper teeth from Bed I agree full well with the primitive type of *C. simum* as described from Laetolil by Dietrich (and earlier by Hilzheimer from Olduvai), and that the dentition of the Bed IV skull is of the same advanced type as that in modern skulls of *C. simum*. The transformation of the Early Pleistocene *Ceratotherium simum germanoafricanum* teeth into those of the modern form was in all probability a gradual process, taking perhaps several hundred thousand years. The material that we have covering this time interval, in the succession of deposits at Olduvai, is extensive, but limb bones and lower teeth as well as fragmentary uppers are of no avail in determining the stage of evolution of the white rhinoceros. Relatively few specimens remain that are characteristic one way or another. It has been shown that two specimens of M³ from Olduvai Bed II are intermediate between that of the Early Pleistocene Baringo white rhino on the one hand, and the modern M³ on the other. One of the Bed II specimens comes from the upper part of Bed II, while the position of the other M³ in Bed II is unknown. Should such specimens be referred to *C. simum germanoafricanum* to *C. simum simum*, or should they be given a new name to designate an intergrading form between the two? I do not consider it expedient to follow the last course. As always in studies of clinal variations in time, the problem arises of how to name the intermediate stages, in so far as they can be recognized in the material available. I believe the best we can do is to indicate the forms transitional between *germano-africanum* and *simum*, if need be, by a bifid racial terminal (cf. Harrison, 1945), thus:

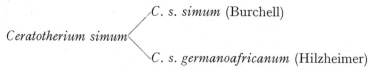

Ceratotherium simum
- *C. s. simum* (Burchell)
- *C. s. germanoafricanum* (Hilzheimer)

Bed I

OLD/60, FLK NN, M¹ or M² sin., damaged anteriorly and at the corners. The medifossette is closed by the union of crochet and crista, the postfossette closed off behind, and the medisinus still open internally. The metaloph is transverse, as in the Laetolil molars figured by Dietrich (1945, Pl. XIII, figs. 1 and 6). The greatest (anterior) basal width of the molar is about 85 mm, at least 10 mm more than that in any recent skull in which this measurement can be taken (Pl. 2, fig. 2).

OLD/63, DK I, 4199, portion of DM² sin.

OLD/63, DK I, III/10, ectoloph fragment.

B.M. (N.H.), M.14805, P² dext. and sin. (1931).

OLD/63, DK I, II/1, 3068, DM₃ dext., length ca. 40 mm.

B.M. (N.H.), M.14808, right mandibular ramus with broken molars.

OLD/41, S.I, F 496, scaphoid sin.

OLD/55, FLK, scaphoid sin., incomplete anteriorly.

OLD/41, S.I, F 520, scaphoid sin., incomplete basally.

OLD/62, DK I, 259, cuneiform dext., incomplete laterally.

OLD/59, KK I, 298, unciform sin.

OLD/62, DK I, 449, unciform dext.

OLD/61, DK I, 28, unciform dext.

OLD/41, S.I, F 861, metacarpal II dext.

OLD/59, MTK I, 97, metacarpal III dext.

OLD/59, MTK I, 98, metacarpal III sin., incomplete proximally.

OLD/41, S.I, F 850, metacarpal III sin., distal end missing.

OLD/62, S.I, a little west of THC, metacarpal III sin., damaged proximally.

OLD/59, FLK, 523, proximal portion of metacarpal III dext.

OLD/62, DK I, 403, metacarpal IV dext.

OLD/61, FLK NN, phalanx I, median digit, length 45 mm, prox. width 63 mm.

OLD/59, KK I, 493, phalanx I, lateral digit, length 41 mm, prox. width 40 mm.

OLD/41, S.I, F 822, phalanx III, median digit, length 26 mm, prox. width 86 mm.

OLD/60, FLK NN, astragalus dext.

OLD/59, FLK, 522, astragalus sin.

OLD/41, S.I, F 486, calcaneum dext., incomplete.

OLD/60, FLK NN, cuboid dext., incomplete.

OLD/60, FLK NN, Tr.II, navicular dext.

OLD/62, FLK NN, Tr.IV, 8849, ectocuneiform sin.

OLD/41, S.I, F 514, ectocuneiform dext.

OLD/52, DK I, 69, metatarsal III dext. without distal end.

Bed I or II

OLD/59, MTK, 105, scaphoid dext.

OLD/41, S.2, 781A, astragalus dext.

OLD/59, FLK I–II, 637, distal portion of lateral metapodial.

Bed II

B.M. (N.H.), M.14810, M³ dext., marked Bed II, has a length of outer surface at base of 78 mm. It is worn to a height of 90 mm from the base, and the occlusal length is 48 mm. The level of 68 mm outer surface length is half way up the crown (45 mm from the occlusal edge), giving a decrease in length of 20 mm over a height

of 45 mm (30 mm in the Baringo, and 10–11 mm in two recent M³, see p. 76).
B.M. (N.H.), DM₄ dext., marked II S, 15/5/35, has a length of 49 mm.

Bed II, Lower part

OLD/62, HWK E II, 581, fragment of right upper molar.

OLD/62, HWK E II, 843, atlas, associated with some other cervical vertebrae.
The atlas is only slightly damaged apart from lacking part of the left wing. The
greatest height is 145 mm, the width over the condyle facets 155 mm. The axis
(no. 804) has an anterior articular facet width of 150 mm and a greatest (posterior)
height of 190 mm; the transverse processes are for the most part missing. The
third cervical vertebra (no. 805) is in two parts: the dorsal arch detached from the
body. Cervicals 4 and 5 (nos. 812 and 799) are in one piece though their processes are
incomplete. The specimens articulate very well and would seem to represent a
single individual. They conform in shape and gradation in characters to those in
recent skeletons.

OLD/54, HWK II, 400, radius dext., without the distal end.

OLD/62, HWK E II, 247, metacarpal III sin.

OLD/62, HWK II, 791, left half of pelvis, with the ilium partially preserved, the
acetabulum entire, and the symphysial border of the obturator foramen missing.
Of the right half (no. 795) the acetabular portion and the ischial branch only are
preserved.

OLD/59, HWK II, 432, distal end of fibula sin.

OLD/–, HWK E II, 820, patella sin.

OLD/62, HWK E II, 234, astragalus dext.

Bed II, Upper part

Diceros bicornis. OLD/63, MNK II I/1, 460, internal portion of an M¹ dext. This
specimen, in the transverse position of the protoloph, absence of medifossette, and
shallow postsinus clearly belongs to the black rhinoceros.

Diceros bicornis. OLD/63, MNK II III/1, 2365, unworn crown of DM₄ sin. Length
45 mm, width at base ca. 23 mm, conforming to its homologue in the black rhino.

Diceros bicornis. B.M. (N.H.), P⁴ dext., marked FLK II S, 2/7/35, anterior width
ca. 65 mm, evidently represents the black rhinoceros.

Diceros bicornis. OLD/52, BK II, 281, anterior outer fragment of left upper molar
the paracone style development of which shows that it represents this species.

OLD/59, FLK II, 705C, P³ sin., incomplete behind.

OLD/59, FLK II, 705D, fragment of left upper molar.

OLD/63, FLK II, Tr.I, ectoloph portion.

OLD/59, FLK, S.23, proximal portion of metacarpal III dext.

OLD/59, FLK, 522, astragalus sin.

OLD/63, FC II, P³ dext., anterior width 63 mm, posterior width 60 mm (Pl. **2** fig. 5).

OLD/63, EF-HR, 211, fragment of right lower molar.

OLD/63, MNK II SK, 140, shaft and distal end of humerus dext.

OLD/63, FC II, 108A, proximal part of humerus sin.

OLD/63, MNK II SK, 1, ulna dext., olecranon incomplete.

OLD/63, MNK II, 3295, calcaneum sin., incomplete.

OLD/57, SHK II, 181, M³ sin., slightly worn, actual height 102 mm externally. The length of the outer surface at base is 71 mm. At 45 mm above the level of 68 mm anteroposterior length (cf. p. 76) the length of the outer surface is 46 mm, giving a decrease of 22 mm (Pl. **5**, fig. 3).

OLD/63, BK II, 1064, M¹ sin., damaged antero-externally. Basal width 73 mm anteriorly, and 60 mm posteriorly (Pl. **2**, fig. 3).

OLD/63, TK II, 2622, M¹ or M² dext., incomplete internally and externally; anterior width ca. 70 mm (Pl. **2**, fig. 4).

OLD/63, TK II, 2833, P³ dext., anterior width 60 mm.

OLD/–, BK II, M³ sin., incomplete behind.

OLD/57, SHK II, 161, central portion of upper molar.

OLD/53, BK II, area C, posterior portion of M³ sin.

OLD/52, BK II, 280, portion of left upper premolar.

OLD/57, BK II, 650, incomplete P² sin., width ca. 40 mm.

OLD/67, BK II, P¹ dext.

OLD/57, BK II, 1015, P³ sin., incomplete antero-externally; width 60 mm.

OLD/57, BK II, 980, P³ dext., anterior width 54 mm.

OLD/52, BK II, 279, P² sin., incomplete; width ca. 40 mm.

OLD/63, TK II, 2075, P² sin., width ca. 40 mm.

OLD/63, BK II, 785, anterior portion of DM³ dext.

OLD/53, BK II Ex., 367, DM³ sin., good specimen. The medifossette and post-fossette pits, exposed at the broken base of the crown, are of the same depth. Although the tooth is worn, the greatest ectoloph length can be taken, and this is within the variation limits of the Makapansgat DM³ of *C. simum*, exceeding that in a recent specimen (Table I).

TABLE I

Measurements of DM³ of *Ceratotherium* (mm)

	Makapansgat	OLD/53, BK II 367	Recent
Greatest length ectoloph	54–61	59	53
Anterior width	48	52	46
Posterior width	46	42	44

OLD/63, BK II, 764, P^2 dext., width ca. 40 mm.

OLD/63, TK II, 2472, DM^1 sin.

OLD/57, SHK II, 112, fragment of upper molar.

OLD/63, BK II, 830, ectoloph portion.

OLD/63, BK II, 630, ectoloph portion.

OLD/63, BK II, area C, fragment of upper molar.

OLD/57, SHK II, S421, medifossette of upper molar.

OLD/53, SHK II, 298, P_{3-4} dext., much worn, in jaw fragment.

OLD/57, BK II, 487, left lower premolar, much worn down.

OLD/57, BK II, 488, M_2 dext., length 54 mm.

OLD/55, BK II, 59, P_3 sin., length 38 mm.

OLD/53, BK II, 366, left lower premolar, worn down to base.

OLD/63, TK II, 2356, incomplete right lower molar.

OLD/57, SHK II, 106, P_1 dext., length 28 mm.

OLD/57, SHK II, 296, fragment of lower molar.

OLD/63, TK II, 2463, fragment of left lower premolar.

OLD/57, SHK II, 245, broken lower molar.

OLD/57, BK II, 1451, juvenile mandible with DM_{1-4} *in situ*.

OLD/63, BK II, 2295, right half of mandible with DM_{2-4}, M_1 erupting. Length from condyle to front 490 mm, height of body at DM_4 130 mm. Tooth measurements hardly differ from those of the foregoing specimen (Table II).

TABLE II

Measurements of lower milk teeth of *Ceratotherium* (mm)

	BK II 1451	BK II 2295
DM_1, ant. post.	25	—
transv.	13	—
DM_2, ant. post.	34	32
transv.	17	17
DM_3, ant. post.	41	38
transv.	22	21
DM_4, ant. post.	48	48
transv.	27	23

OLD/63, TK II, 2452, proximal portion of scapula dext.

OLD/53, BK II Ex., 438, proximal portion of scapula sin.

OLD/63, BK II, 3122, proximal portion of scapula dext., utilized.

OLD/63, BK II, 1974, humerus dext., incomplete at either end.

OLD/63, BK II, 3014B, medial distal condyle of humerus dext.

OLD/53, BK II, Ex., proximal portion of radius dext.

OLD/55, BK II, 127, unciform dext., posterior process incomplete.

OLD/57, SHK S, 95, phalanx I, median digit, length 34 mm, prox. width 44 mm.

OLD/52, BK II, 324, phalanx II, median digit, length 27 mm, prox. width 41 mm.

OLD/53, SHK II, 298, phalanx III, median digit, length 23 mm, prox. width 64 mm.

OLD/52, BK II, 323, phalanx I, lateral digit, length 32 mm, prox. width 34 mm.

OLD/52, BK II, 325, phalanx I, lateral digit, length 29 mm, prox. width 27 mm.

OLD/63, BK II, 878, tibia sin. without proximal end.

OLD/52, SHK II, 692, astragalus dext.

OLD/63, TK II, 2492, astragalus dext.

OLD/53, SHK II, 287, astragalus dext.

OLD/57, BK II, 661, astragalus dext.

OLD/63, TK II, 2407, calcaneum sin.

OLD/55, BK II, 22, cuboid dext.

OLD/53, BK II Ex., 131, navicular sin.

OLD/57, BK II, 520, ectocuneiform dext.

OLD/53, SHK II, 288, ectocuneiform sin.

Bed III

Diceros bicornis. B.M. (N.H.), M.14799, M³ dext., is one of the few specimens from Bed III belonging to this species. It measures 48 mm anteroposteriorly (on the inner side); the anterotransverse diameter is 58 mm, and the length of the outer surface 57 mm. It is indistinguishable from recent M³ of *D. bicornis*, but as it is worn the crown height cannot be determined.

In the collection at the National Museum Centre for Prehistory, Nairobi, there are rhinoceros specimens from JK2, Geol. Pit 8, an excavation made by Miss Dr. Maxine Kleindienst (now Mrs. Haldemann-Kleindienst), to whom I am indebted for information relating to the stratigraphical distribution of the finds (in litt., July 29, 1967). As noted in a published report (Kleindienst, 1964) various Beds are exposed, but the material from Trench 8 is the only excavated material from Bed III that can be considered to belong to that formation and not to be derived. One of the most interesting specimens of this dig is an upper milk molar of the common species, *Ceratotherium simum*, larger than a recent specimen but not quite as large as its homologues from the Makapansgat Limeworks Cave in South Africa (Hooijer, 1959): OLD/62, JK2, Geol. Pit. 8, section 5, no. 1618, DM² dext. The parastyle is rather raised, and there is no cingulum along the internal face but only a small tubercle at the medisinus entrances, springing from the base of the hypocone. The crista is bifid, and one arm joins the crochet, forming a medifossette.

OLD/62, JK2, Geol. Pit 8, section 6, no. 1593, P₃ dext., length 37 mm.

OLD/62, JK2, Geol. Pit 8, section 1, no. cem. 178, humerus dext., slightly damaged proximally.

OLD/62, JK2, Geol. Pit 8, no. 379, unciform dext.

TABLE III

Measurements of DM² of *Ceratotherium* (mm)

	Makapansgat	OLD/62, JK2 1618	Recent
Greatest length ectoloph	42–51	42	41
Anterior width	39–41	39	36
Posterior width	39–43	38	35

OLD/62, JK2, Geol. Pit 8, section 5, no. 1612, acetabular portion of left innominate. OLD/62, JK2, Geol. Pit 8, no. 1961, proximal end of metatarsal III sin. Some specimens in B.M. (N.H.), marked Bed III, also represent the common *Ceratotherium*: M.14807, left mandibular ramus with M_{2-3} and right ramus with P_2 (broken) through M_3, height at M_1 130 mm; M.14809, right ramus of the mandible with broken molars, and (unnumbered) an M_3 sin., 17/5/35, with a length of 65 mm.

Bed III or IV

Here belong various specimens from JK2, Trenches A and B, which are either from Bed III or from Lower Bed IV, excavated by Miss Kleindienst in 1962:

Diceros bicornis. OLD/62, JK2/A, no. 1025, DM² dext. It lacks the outer portion but shows the continuous internal cingulum by which it is distinguished from its homologue in *Ceratotherium*. No medifossette is formed as there is no crista. Unfortunately no measurements can be given (Pl. **3,** fig. 6).

Diceros bicornis. OLD/62, JK2/B, floor, square M4, no. 13, central fragment of left upper molar. The small crochet does not extend across the medisinus, and part of the postsinus is present, showing it to be much shallower than the medisinus. These are characters that stamp this specimen, incomplete as it is, as belonging to the black rhinoceros.

OLD/62, JK2/A, no. 2806, P³ sin., incomplete internally; width ca. 58 mm.

OLD/62, JK2/A, no. 3269, upper portion of P³ sin., incomplete behind (Pl. **3,** figs.1–2).

OLD/62, JK2/A, no. 3172, ectoloph portion.

OLD/62, JK2/A, no. 3056, ectoloph portion.

OLD/62, JK2/A, no. 2446, M_{1-3} in ramus fragment; length of M_3 at alveolus 65 mm.

OLD/62, JK2/B, floor N4–21, M_2 dext., length 54 mm.

OLD/62, JK2/B, P_2 dext., length 35 mm.

OLD/62, JK2/A, no. 832, left lower premolar, much worn down.

OLD/62, JK2/A, no. 1687, magnum dext., downward process incomplete.

OLD/62, JK2/B, no. 1127, metacarpal III dext., incomplete at both ends.

OLD/62, JK2/A, no. 484, base of trochanter tertius of femur dext.

There are further some *Ceratotherium* specimens in the 1961 Olduvai collection marked JK2/B, including an M_1 dext., length 50 mm, and some molar fragments with the characteristic medifossettes and parastyles.

Bed IV

Diceros bicornis. B.M. (N.H.), CMK IV S, 20/5/35, is an M^3 sin. displaying all the distinctive characters of the modern black rhinoceros.

B.M. (N.H.), M.14812, P^3 sin., marked IV '31; width 50 mm.

OLD/55, BK II, 95, originating from Bed IV, is a damaged M^3 sin. The crown was probably unworn; the greatest height preserved is 85 mm. The anteroposterior length of the outer surface at base is 68 mm, and that 45 mm higher up is 63 mm, giving a decrease of only 5 mm (30 mm in the Baringo, 20–22 mm in two Bed II, and 10–11 mm in two recent M^3, see p. 76).

OLD/52, MRC, 641, M_2 sin., much worn down.

OLD/62, Rhino Korongo, the greater part of the left half of a skull, together with an entire mandible. The nasal bones and the top of the occiput are missing as well as most of the palate and the skull base apart from the left inferior squamosal processes and the occipital condyles. The full left cheek teeth series P^2–M^3 is preserved (Pl. **6,** figs. 1–2). There is nothing in the structure of the teeth that distinguishes them from those in modern *C. simum*: the transverse lophs of the P and M are as obliquely placed as those in recent skulls. Skull characters likewise are those of the modern white rhinoceros, but the skull is impressive because of its great size (Table VIII). The condylo-basal length, which can be exactly taken, exceeds that in all of the modern *C. simum* skulls recorded by Heller (1913: maximum 750 mm), and the total length of the mandible is greater than in all but one of Heller's specimen's, viz., the old male from South Africa that has the maximum skull length: its mandibular length is 635 mm.

To the same skull belongs the right M^2 with a portion of the palate, showing the lateral palatine foramen and part of the posterior palatine notch (Pl. **3,** fig. 8). The right M^2 differs from the left M^2 of the skull in that the medifossette is not closed off internally but remains confluent with the medisinus, an individual aberration occasionally seen in recent molars as well.

LAETOLIL, TANZANIA

All the rhinoceros specimens from this site that I have seen belong to *Ceratotherium simum*, and some of the best specimens show the differential characters on which Dietrich (1942, 1945) based his *Serengeticeros efficax*.

In the 1959 Laetolil collection at the Nairobi Centre there is a DM^3 sin. (no. 116), much worn down (Pl. **3,** fig. 7) measuring 48 mm anterotransversely and 46 mm posterotransversely, as do Makapansgat specimens (Hooijer, 1959). An incomplete DM³ dext. (no. 119) is 50 mm wide anteriorly. A P³ sin. (no. 582) is much worn, and incomplete internally; the only enamel pit left on the occlusal surface is that of the medisinus. The anterior width of another P³ sin. (no. 418) is 51 mm; that of a P⁴ sin. (no. 346), 61 mm, the same as that in the anterior half of a left tooth (no. 439) that may represent P⁴.

A P³ and P⁴ dext. in the 1935 Laetolil collection in the British Museum (Natural History) both incomplete anteroexternally, show the metaloph to be transversely placed, and not oblique as in the modern white rhino. The P³ is 61 mm transversely at base, the P⁴, 66 mm. Unfortunately, the height is not known in any of the Laetolil teeth as they are all worn.

A right M³ in the Nairobi Centre collection (no. 194) lacks all the corners except for the anterointernal one; hence, no measurements can be given.

There is, in the Nairobi collection, a mandible from Laetolil in a rather fragmentary state; the symphysial portion with the much worn crowns of the premolars, and further part of the right ramus with root stumps of some of the molars. In the British Museum (Natural History) collection there is a left mandible with P_3–M_3 (length 235 mm); the height at M_1 is only 100 mm, and the symphysis is broken.

OMO BASIN, S. ETHIOPIA

A well-preserved skull of *Ceratotherium simum* from the Upper Series of the Omo Basin (with *Loxodonta africana* and *Phacochoerus aethiopicus*) (AP 671–1), found at a site called " Rhino Canyon ", was sent to Nairobi through the courtesy of Professor Clark Howell in July 1967. It is stated to come from probably near the known lower limits of the Upper Series of the Omo Basin. The skull (Pl. **4,** figs. 1–2) is laterally crushed, and lacks the tip of the nasal bones. The full left dentition P²–M³ is *in situ* as well as the right P³–M³. The last molar is just touched by wear. From the few measurements that can be given (Table VIII) the specimen does not appear to be particularly large, less so than the largest in Heller's series from South Africa. The tooth patterns conform to those in the modern white rhino.

An isolated P_3 sin. of *C. simum*, measuring 48 by 28 mm, is no. 70 of the collection from the Omo Upper Level made by R. Leakey in 1967.

A palatal portion of the skull, with M²⁻³ sin., and a broken P⁴ sin. (no. 24) of the Omo 1967 collection made by R. Leakey, is from the lower level, and shows the Early Pleistocene features (Pl. **5,** figs. 4–5).

The following specimens of *Ceratotherium simum* originate from the White Sands of the Omo Basin (cf. Howell, 1968):

no. 591, unworn crown of P³ sin., base incomplete. Greatest length of ectoloph 58 mm, height over 85 mm (Pl. **3,** figs. 4–5).

no. 333, M_3 sin., length 64 mm, width 35 mm.

no. 492, right lower molar, length 57 mm, height of posterior loph (barely worn), 78 mm.

no. 409, left lower molar, incomplete in front.

no. 332, P_3 dext., much worn, and

no. 549, central fragment of left upper molar.

There is an astragalus sin., no. 491, slightly damaged, from the Omo White Sands; measurements in Table XXIII.

All the remaining Omo rhino teeth collected by the 1967 parties listed below belong to *Diceros bicornis*, a species not recorded before from Omo. There are two unworn specimens of M^3 that show the most distinctive character of the fossil *D. bicornis* as compared with the modern form. The outer surface of an M^3 sin. from the White Sands (no. 12; Pl. **5,** fig. 1), as well as an M^3 dext. collected by the French party, have the height just slightly greater than the basal length of the outer surface. The French specimen comes from a level provisionally regarded as Upper Villafranchian (Arambourg *et al.*, 1967, p. 1895, as *Diceros* cf. *bicornis*), and the American specimen is Early Pleistocene as well.

TABLE IV

Measurements of M^3 of *Diceros bicornis* (mm)

	Omo sin. unworn	Omo dext. unworn	Nairobi Recent unworn	Nairobi Recent worn
Ant. post. (internally)	—	50	50	49
Anterotransverse	—	52	51	55
Length of outer surface	55	58	54	60
Height of ectoloph	56	59	64	—

In the modern *D. bicornis* the unworn M^3 has the outer surface distinctly higher than wide; in a specimen recorded by Cooper (1934, p. 581) the height from the level between the roots to the peak of the crown is 84 mm, and the maximum breadth 71 mm. An unworn M^3 in the osteological collection at the Nairobi Centre (no. 6)

TABLE V

Measurements of DM^3 of *Diceros bicornis* (mm)

	Omo	Makapansgat	Recent
Greatest length ectoloph	47	47–52	45–49
Anterior width	45	46–60	40–49
Posterior width	42	43–47	37–43

has the height of the ectoloph exceeding the basal length by 10 mm (Table IV). An outer view of the recent specimen is given in Pl. **5,** fig. 2.

A specimen in the Omo collection of 1967 made by R. Leakey is a DM³ sin. (no. 135) from the Lower Level, which is on the small side compared with its homologues of *Diceros bicornis* from Makapansgat (Hooijer, 1959).

Apart from the outer surface of the unworn M³ sin. there are two more specimens of *Diceros bicornis* originating from the Omo White Sands, viz.,

no. 461, inner portion of protoloph and medisinus of P³ sin. with heavy internal cingulum, and

no. 598, M₃ sin., length 57 mm, width 36 mm.

KANAM WEST, KENYA

From Kanam West there are rhinoceros teeth representing the two recent species. This site is considered Early Pleistocene.

In the British Museum (Natural History) collection there is an M³ dext. of *Ceratotherium simum*, M.15888, which measures 84 mm at the base of the outer surface. The height of the slightly worn crown is just over 100 mm. The outer surface is 74 mm in length half way up the crown, and then reduced to 50 mm at the top due to the convexity of the anterior edge. This specimen is rather higher than Dietrich (1945, p. 59) would admit for the Early Pleistocene form from Laetolil, but close to an Omo M³ of *C. simum germanoafricanum* (Arambourg, 1947, p. 197).

M.15892 is an upper dentition of *Diceros bicornis* from Kanam West, embedded in plaster and lacking only the left M³. They represent an old individual, and their dimensions do not exceed those in the Recent or the Upper Pleistocene Hopefield *Diceros bicornis* specimens (Hooijer and Singer, 1960).

TABLE VI

Measurements of upper teeth of *D. bicornis* from Kanam West (mm)

P², ant. transv.	38	M¹, ant. post	ca. 47
post. transv.	41	ant. transv.	63
P³, ant. post.	34	post. transv.	60
ant. transv.	52	M², ant. transv.	62
post. transv.	—	post. transv.	55
P⁴, ant. post.	40	M³, ant. post.	49
ant. transv.	58	ant. transv.	56
post. transv.	56	length outer surface	57

In the Nairobi Centre collection there is a DM⁴ dext. from Kanam West, F.3516, a surface find (Pl. **3,** fig. 3) which is entire and about half worn down (the greatest length of the ectoloph, therefore, cannot be taken). It is similar in size to the

Makapansgat DM^4 of *D. bicornis* (Hooijer, 1959), and somewhat larger than its homologue in extant *D. bicornis* (Table VII).

TABLE VII

Measurements of DM^4 of *Diceros bicornis* (mm)

	Kanam	Makapansgat	Recent
Ant. post.	45+	—	49–55
Ant. transv.	54	52–53	45–52
Post. transv.	49	51	40–47

There is further a worn DM^4 sin., with the antero-external angle missing, with a postero-transverse diameter of likewise 49 mm, and two upper molar fragments (F.3520 and F.2517).

NAIVASHA

In the collection at the National Museum Centre for Prehistory and Palaeontology in Nairobi there is a mandible of *Ceratotherium simum* found by Mr. J. K. Ker on his farm near Knights, about 10 miles from Naivasha, at a depth of four feet in white soil, on top of an escarpment. The specimen, brought in by Wakeford Thompson, is hardly fossilized in appearance. It is the only white rhinoceros specimen thus far from this site (Naivasha is some 50 miles N.W. of Nairobi) and determination of its geological age would be of considerable interest.

The symphysial portion is incomplete and the ascending rami missing; the teeth present are P$_3$–M$_2$ on either side, and the right M$_3$. The mental foramen is below the P$_3$–P$_4$ junction. The premolars have both valleys closed off from the lingual margin; M$_1$ has the anterior valley worn out and only the posterior remaining. M$_2$ has the enamel figures of metalophid and hypolophid just joining occlusally, while M$_3$ is only touched by wear. The M$_3$ is 66 mm long by a height of ca. 70 mm. The teeth do not differ in size from Recent specimens and the stage of wear corresponds with that of the old female mandibles nos 8 and 9 on Plate **28** of Heller (1913).

OLORGESAILIE

An entire tibia dext., marked Olorgesailie, site 10, Basal Beds B (surface) August 1944. Measurements in Table XXII.

7*

TABLE VIII

Measurements of skull and mandible of *Ceratotherium simum* subsp. (mm)

	Leiden Museum cat. b	Nairobi Centre no. 34	Olduvai Bed IV	Baringo J.M.91	Omo AP671–1
Occipito-nasal length	780	825	ca.920	—	—
Condylo-basal length	700	720	780	—	ca. 700
Zygomatic width	345	340	—	ca. 380	—
Lacrymal width	315	280	—	—	—
Postorbital constriction	120	115	—	—	—
Width at occipital crest	225	220	—	—	—
Bicondylar width	165	155	175	160	155
Depth of dorsal concavity	55	45	—	—	—
Least depth of zygoma	80	70	—	—	—
From foramen magnum to occipital crest	170	170	—	—	ca. 180
Width of nasal boss	160	195	—	—	—
Separation inferior squamosal processes	5	10	10	—	—
Length P^2–M^2	210	260	270	—	250
Mandible, total length	595	585	630	—	—
Length of symphysis	155	135	ca. 165	—	—
Width at symphysis	115	105	ca. 130	—	—
Height of body at M_1	125	115	140	—	—
Width of ramus at angle	160	155	165	—	—
Length P_2–M_2	190	250	275	—	—

TABLE IX

Measurements of scapula (mm)

	bicornis	*simum*	TK II 2452	BK II 438	BK II 3122
Ant. post. diam. of glenoid cavity	85	100	110	110	ca. 110
Transv. diam. of idem	80	95	—	—	95
Transv. diam. tuber scapulae	45	55	75	60	—

TABLE X

Measurements of humerus (mm)

	bicornis	simum	BK II 1974	MNK II 140	JK2 178
Length, caput to medial condyle	350	400	380	—	350
Width over caput and post. part of lateral tuberosity	145	180	—	—	ca. 155
Width at deltoid tuberosity	130	170	—	—	140
Least width of shaft	60	85	80	75	70
Greatest distal width	150	180	—	—	150
Width of trochlea	100	120	110	105	100

TABLE XI

Measurements of radius (mm)

	bicornis	simum	HWK II 400	BK II
Median length	345	380	—	—
Proximal width	100	120	115	ca.100
Mid-shaft width	55	70	65	—
Distal width	95	120	—	—

TABLE XII

Measurements of ulna (mm)

	bicornis	simum	MNK II
Greatest length	450	510	—
Width at semilunar notch	90	110	90
Mid-shaft width	45	60	50
Greatest distal width	75	90	70

TABLE XIII
Measurements of scaphoid (mm)

	bicornis	simum	S.I. F 496	MTK 105	FLK	S.I. F 520
Posterior height	50	62	65	ca. 65	60	60
Anterior height	60	65	70	60	—	—
Proximal width	55	60	75	70	—	—
Max. diam. distal facets	70	73	85	83	ca. 70	—

TABLE XIV
Measurements of lunar (mm)

	bicornis	simum
Anterior height	48	60
Proximal width	48	58
Max. ant. post. diameter	68	75

TABLE XV
Measurements of cuneiform (mm)

	bicornis	simum	DK I 259	F 836*
Anterior height	50	58	70	52
Distal width	38	45	—	43
Prox. ant. post. diameter	40	48	ca. 55	48
Max. horizontal diameter	53	66	ca. 80	57

* Olduvai, no site or level given, but entire (right) specimen.

TABLE XVI
Measurements of magnum (mm)

	bicornis	simum	JK2 1687
Anterior height	32	38	34
Anterior width	49	58	ca. 45
Prox. ant. post. diameter	67	70	67
Greatest diameter	85	84	90

TABLE XVII
Measurements of unciform (mm)

	bicornis	simum	KK I 298	BK II 127	DK I 449	DK I 28	JK2 379
Anterior height	51	55	66	54	68	73	56
Greatest width	63	74	ca. 90	74	97	110	70
Greatest ant. post. diam.	90	99	ca. 110	—	122	134	93

TABLE XVIII
Measurements of metacarpals (mm)

Mc. II	bicornis	simum	S.I F 861
Median length	147	160	183
Proximal width	32	45	47
Proximal ant. post. diameter	46	49	58
Middle width	33	40	44
Middle ant. post. diameter	19	24	30
Greatest distal width	39	50	53
Width of distal trochlea	33	40	48
Distal ant. post. diameter	41	45	58
Ratio middle width/length	0·22	0·25	0·24

Mc. III	bicornis	simum	HWK II 247	S.I w.THC	MTK 97	I 98	S.I 850	FLK 523
Median length	162	176	193	210	205	208	—	—
Proximal width	59	68	86	—	84	—	ca.90	70
Proximal ant. post. diameter	48	52	66	—	66	—	70	ca. 70
Middle width	46	58	66	68	64	65	67	—
Middle ant. post. diameter	22	28	34	34	33	32	33	—
Greatest distal width	61	71	84	—	83	86	—	—
Width of distal trochlea	51	59	71	—	66	68	—	—
Distal ant. post. diameter	44	48	54	59	57	60	—	—
Ratio middle width/length	0·28	0·33	0·34	0·32	0·31	0·31	—	—

TABLE XVIII—(*continued*)

Mc. IV	bicornis	simum	DK I 403
Median length	136	143	166
Proximal width	43	54	65
Proximal ant. post. diameter	43	50	54
Middle width	33	41	48
Middle ant. post. diameter	18	23	31
Greatest distal width	43	52	59
Width of distal trochlea	37	42	52
Distal ant. post. diameter	38	45	51
Ratio middle width/length	0·24	0·29	0·29

TABLE XIX

Measurements of pelvis (mm)

	bicornis	simum	HWK II 791	JK2 1612
Diameter of acetabulum	90	115	120	ca. 95
Greater diam. of obturator for.	105	115	125	—
Least width of ilium shaft	70	90	110	—

TABLE XX

Measurements of femur (mm)

	bicornis	simum
Greatest length	440	530
Proximal width	195	225
Least width of shaft	60	80
Greatest distal width	120	150
Distal ant. post. diameter, medial side	160	190
Diameter of caput	80	95

TABLE XXI

Measurements of patella (mm)

	bicornis	simum	HWK II 820	OLD*
Length	95	107	107	125
Width	87	93	105	110

*Olduvai, no site or level given, but particularly large specimen.

TABLE XXII

Measurements of tibia (mm)

	bicornis	simum	BK II 878	F 3312*	F 3313*	Olorgesailie
Greatest length	335	380	—	—	—	370
Proximal width	110	135	—	—	—	115
Distal width	85	95	85	110	90	90
Distal ant. post. diameter	70	80	70	85	70	67

* Two OLD/41 specimens, marked S 4 and S 2, respectively, both distal ends of tibiae sin.

TABLE XXIII

Measurements of astragalus (mm)

	bicornis	simum	HWK II 234	FLK NN	S2 781A	SHK II 692
Lateral height	71	74	100	91	88	84
Medial height	70	75	99	88	90	84
Total width	83	95	118	108	109	106
Ratio medial height/total width	0·84	0·79	0·84	0·81	0·83	0·79
Trochlea width	78	83	108	101	93	96
Width of distal facets	72	85	103	93	90	86

	TK II 2429	SHK II 287	BK II 661	FLK 522	Omo White Sands
Lateral height	—	73	ca. 75	—	66
Medial height	90	72	79	ca. 95	65
Total width	108	82	99	ca. 115	74
Ratio medial height/total width	0·83	0·88	0·79	ca.0·83	0·88

TABLE XXIV

Measurements of calcaneum (mm)

	bicornis	simum	TK II 2407	S.I F 486	MNK II 3295
Lateral height	110	125	140	ca. 140	120
Greatest width	70	80	95	—	—
Ant. post. diameter	65	75	80	80	70

D. A. HOOIJER

TABLE XXV
Measurements of cuboid (mm)

	bicornis	simum	BK II 22	FLK NN	OLD/51* 403
Anterior height	37	43	46	50	46
Anterior width	44	52	46	—	42
Greatest ant. post. diameter	65	80	80	—	75

* No site or level given: entire cuboid sin.

TABLE XXVI
Measurements of navicular (mm)

	bicornis	simum	FLK NN Tr. II	BK II 131
Anterior height	24	29	31	23
Total width	45	55	—	45
Ant. post. diameter	56	62	64	52

TABLE XXVII
Measurements of ectocuneiform (mm)

	bicornis	simum	FLK NN 8894	S.I F 514	SHK II 288	BK II 520
Anterior height	24	27	31	31	25	28
Anterior width	45	57	67	70	43	57
Ant. post. diameter	53	54	66	65	46	56

TABLE XXVIII
Measurements of metatarsals (mm)

Mt. II	bicornis	simum
Median length	129	151
Proximal width	25	34
Proximal ant. post. diam.	42	47
Middle width	25	28
Middle ant. post. diam.	19	24
Greatest distal width	33	39
Width of distal trochlea	29	36
Distal ant. post. diameter	36	40
Ratio middle width/length	0·20	0·19

TABLE XXVIII—(*continued*)

Mt. III	*bicornis*	*simum*	DK I 69	JK2 1961
Median length	148	169	—	—
Proximal width	48	55	74	57
Proximal ant. post. diameter	48	49	ca. 55	52
Middle width	40	48	67	—
Middle ant. post. diameter	21	25	38	—
Greatest distal width	54	66	—	—
Width of distal trochlea	47	51	—	—
Distal ant. post. diameter	42	47	—	—
Ratio middle width/length	0·27	0·28	—	—

Mt. IV	*bicornis*	*simum*
Median length	125	146
Proximal width	42	49
Proximal ant. post. diameter	40	45
Middle width	26	29
Middle ant. post. diameter	24	28
Greatest distal width	36	39
Width of distal trochlea	33	37
Distal ant. post. diameter	38	41
Ratio middle width/length	0·21	0·20

REFERENCES

ARAMBOURG, C. 1938. Mammifères fossiles du Maroc. *Mém. Soc. Sci. nat. Maroc*, **46**: 1–74, 9 pls., 15 figs.

—— 1947. Contribution à l'étude géologique et paléontologique du Lac Rodolphe et de la basse vallée de l'Omo, part 2, Paléontologie. In *Mission Scientifique de l'Omo 1932–1933*, Vol. 1, fasc. 3, 231–562, 40 pls., 91 figs.

ARAMBOURG, C., CHAVAILLON, J. and COPPENS, Y. 1967. Premiers résultats de la nouvelle Mission de l'Omo (1967). *C.R. hebd Séanc Acad. Sci., Paris* **265**: ser. D, 1891–96, 1 fig.

COOKE, H. B. S. 1950. A critical revision of the Quaternary Perissodactyla of southern Africa. *Ann. S. Afr. Mus.* **31**: 393–479, 31 figs.

COOPER, C. F. 1934. The extinct rhinoceroses of Baluchistan. *Phil. Trans. R. Soc.* B, **223**: 569–616, pls. 64–67, 21 figs.

DIETRICH, W. O. 1942. Zur Entwicklungsmechanik des Gebisses der afrikanischen Nashörner. *Zentbl. Miner*, B, 297–300, 3 figs.

—— 1945. Nashornreste aus dem Quartär Deutsch-Ostafrikas. *Palaeontographica* **96** A: 46–90, pls. XIII–XIX.

GROVES, C. P. 1967. Geographic variation in the Black Rhinoceros, Diceros bicornis (L., 1758). *Z. Säugetierk.* **32**: 267–276, 2 figs.

GUGGISBERG, C. A. W. 1966. *S.O.S. Rhino*. A survival book. Nairobi (East African Publishing House), 174 pp., pls., figs.

HARRISON, J. M.   1945.   Races, Intermediates and Nomenclature—a suggested Modification of the Trinomial System.   *The Ibis* **87**: 48–51.

HELLER, E.   1913.   *The White Rhinoceros.*   Smithson. Misc. Coll., Vol. 61, no. 1, 77 pp., 31 pls.

HILZHEIMER, M.   1925.   Rhinoceros simus germano-africanus n.sub-sp. aus Oldoway. Wiss. Erg. Oldoway-Expedition 1913, N.F., Heft 2, 45–79, 1 pl., 4 figs.

HOOIJER, D. A.   1950.   The Study of Subspecific Advance in the Quaternary.   *Evolution* **4**: 360–361.

—— 1959.   Fossil Rhinoceroses from the Limeworks Cave, Makapansgat.   *Palaeont. afr.* **6**: 1–13, 4 figs.

—— 1966.   Miocene rhinoceroses of East Africa.   *Bull. Br. Mus. nat. Hist., Geol.,* **13**: 2, 117–190, pls. 1–15.

—— 1968.   A rhinoceros from the Late Miocene of Fort Ternan, Kenya.   *Zoöl. Meded., Leiden,* **43**: no. 6, 77–92, 3 pls.

HOOIJER, D. A. and SINGER, R.   1960.   Fossil rhinoceroses from Hopefield, South Africa.   *Zoöl. Meded., Leiden,* **37**: no. 8, 113–128, pl. XI.

HOPWOOD, A. T.   1926.   Fossil Mammalia.   In *The Geology and Palaeontology of the Kaiso Bone Beds.*   Occ. Papers Geol. Surv. Uganda Protect., no. 2, 13–36, 3 pls., 14 figs.

HOWELL, C.   1968.   Omo Research Expedition.   *Nature, Lond.* **219**: no. 5154 (Aug. 10), 567–572, 3 figs.

KLEINDIENST, M. R.   1964.   Summary report on excavations at site JK2, Olduvai Gorge, Tanganyika, 1961–1962.   Ann. Report Antiquities Division for 1962, Tanganyika Ministry of National Culture and Youth, Dar es Salaam (1964), pp. 4–6, 2 figs.

LEAKEY, L. S. B.   1965.   *Olduvai Gorge 1951–1961.*   Vol. I, A preliminary report on the geology and fauna.   Cambridge University Press, XIV + 118 pp., 97 pls., 2 figs., 2 maps, frontispiece.

LEAKEY, M. D.   1965.   Descriptive list of the named localities in Olduvai Gorge, In.   *Olduvai Gorge 1951–1961* ed. L. S. B. LEAKEY, pp. 101–109, sketch-map.

MARTYN, J. and TOBIAS, P. V.   1967.   Pleistocene deposits and new fossil localities in Kenya.   *Nature, Lond.* **215**: no. 5100 (July 29), 476–480, 4 figs.

THENIUS, E.   1955.   Zur Kenntniss der unterpliozänen Diceros-Arten (Mammalia, Rhinocerotidae).   *Ann. naturh. Mus. Wien* **60**: 202–211, 6 figs.

ZEUNER, F. E.   1934.   Die Beziehungen zwischen Schädelform und Lebensweise bei den rezenten und fossilen Nashörnern.   *Ber. naturf. Ges. Freiburg. i. B.* **34**: 21–80, 8 pls.

LEGENDS TO PLATES

PLATE 1

Ceratotherium simum germanoafricanum (Hilzheimer)

FIG. 1. Skull, Chemeron Formation, Lake Baringo, J.M.91, palatal view. × 1/5.
FIG. 2. Same, right view. × 1/5.

PLATE 2

Ceratotherium simum germanoafricanum (Hilzheimer)

FIG. 1. Maxillary with M^{1-3} sin., Chemeron Formation, Lake Baringo, J.M.507, crown view. × 2/3.
FIG. 2. M^1 or M^2 sin., OLD/60, FLK NN, crown view. × 2/3.

Ceratotherium simum (Burchel) subsp.

FIG. 3. M^1 sin., OLD/63, BK II, 1064, crown view. × 2/3.
FIG. 4. M^1 or M^2 dext., OLD/63, TK II, 2622, crown view. × 2/3.
FIG. 5. P^3 dext., OLD/63, FC II, crown view. × 2/3.

PLATE 3

Ceratotherium simum (Burchell) subsp.

FIG. 1. P^3 sin., OLD/62, JK2/A, no. 3269, crown view. × 2/3.
FIG. 2. Same, outer view. × 2/3.
FIG. 4. P^3 sin., Omo White Sands, 1967, 591, crown view. × 2/3.
FIG. 5. Same, outer view. × 2/3.
FIG. 7. DM^3 sin., Laetolil, 1959, 116, crown view. × 2/3.
FIG. 8. M^2 dext., OLD/62, Rhino Korongo, crown view. × 2/3.

Diceros bicornis (L.) subsp.

FIG. 3. DM^4 dext., Kanam West, F.3516, crown view. × 2/3.
FIG. 6. DM^2 dext., OLD/62, JK2/A, no. 1025, crown view. × 2/3.

PLATE 4

Ceratotherium simum (Burchell) subsp.

FIG. 1. Skull, Omo, Upper Series, 1967, Rhino Canyon, palatal view. × 1/5.
FIG. 2. Same, right view. × 1/5.

PLATE 5

Diceros bicornis (L.) subsp.

FIG. 1. M^3 sin., Omo White Sands, 1967, 12, outer view. × 2/3.
FIG. 2. M^3 sin., Recent, Dept. of Osteology, Nairobi, no. 6, outer view. × 2/3.

Ceratotherium simum (Burchell) subp.

FIG. 3. M^3 sin., OLD/57, SHK II, 181, outer view. × 2/3.

Ceratotherium simum germanoafricanum (Hilzheimer)

FIG. 4. Palatal portion of skull with M^{2-3} sin., Omo, Lower Level, 1967, crown view. × 1/3.
FIG. 5. Skull fragment with broken P^4 sin., Omo, Lower Level, 1967, no. 24, crown view. × 2/3.

PLATE 6

Ceratotherium simum (Burchell) subsp.

FIG. 1. Skull, OLD/62, Rhino Korongo, left view. × 1/5.
FIG. 2. P^2–M^3 sin. of same skull, crown view. × 1/2.

PLATE I

1

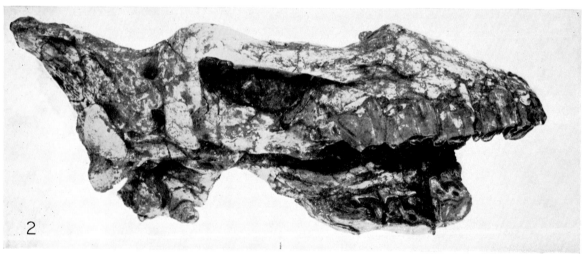

2

PLATE 2

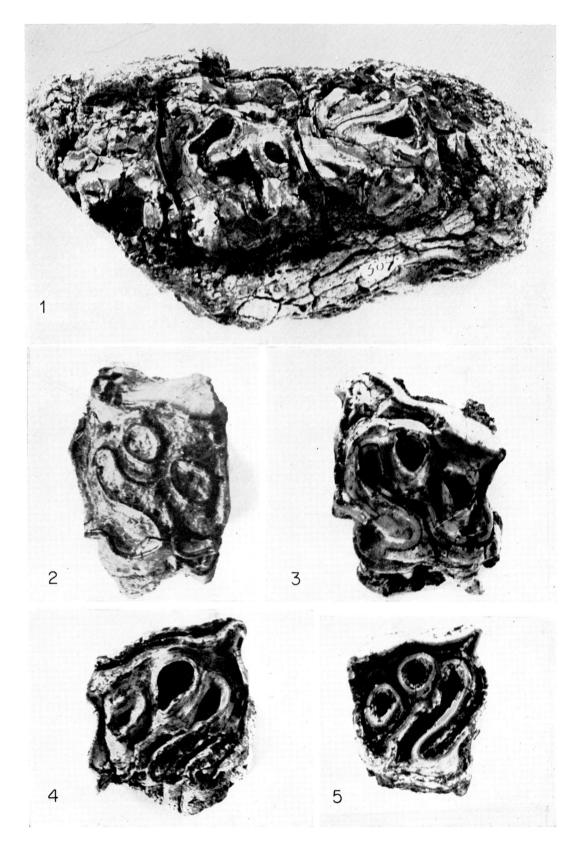

PLATE 3

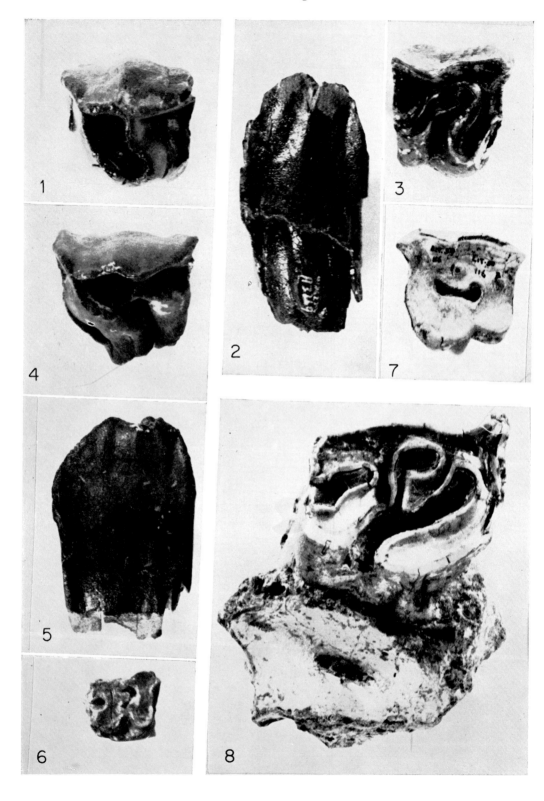

PLATE 4

1

2

PLATE 5

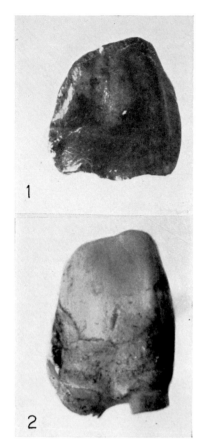

1

2

4

3

5

PLATE 6

1

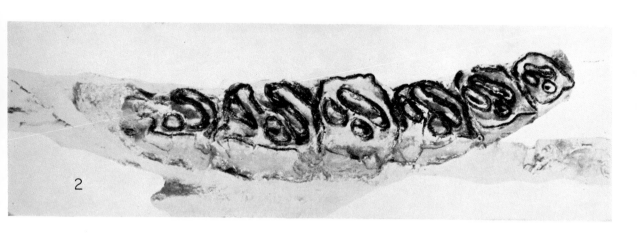

2

AUTHOR INDEX

The numbers in *italics* indicate the pages on which
names are mentioned in the reference list.

Fossil Mammals of Africa

Series 1

No. 1. " The Miocene Hominoidea of East Africa " by W. E. Le Gros Clark and L. S. B. Leakey. 1951, £1 15s.

No. 2. " The Pleistocene Fauna of Two Blue Nile Sites " by A. J. Arkell, D. M. A. Bate, L. H. Wells and A. D. Lacaille. 1951, 15s.

No. 3. " Associated Jaws and Limb Bones of Limnopithecus macinnesi " by W. E. Le Gros Clark and D. P. Thomas. 1951, 15s.

No. 4. " Miocene Anthracotheriidæ from East Africa " by D. G. MacInnes. 1951, 12s 6d.

No. 5. " The Miocene Lemuroids of East Africa " by W. E. Le Gros Clark and D. P. Thomas. 1952, 12s 6d.

No. 6. " The Miocene and Pleistocene Lagomorpha of East Africa " by D. G. MacInnes. 1953, 10s.

No. 7. " The Miocene Hyracoids of East Africa " by T. Whitworth. 1954, £1 5s.

No. 8. " An Annotated Bibliography of the Fossil Mammals of Africa " by A. Tindell Hopwood and J. P. Hollyfield. 1954, £2 5s.

No. 9. " A Miocene Lemuroid Skull from East Africa " by Sir Wilfred Le Gros Clark. 1956, 5s.

No. 10. " Fossil Tubulidentata from East Africa " by D. G. MacInnes. 1956, £1.

No. 11. " Erinaceidæ from the Miocene of East Africa " by P. M. Butler. 1957, £2.

No. 12. " A New Miocene Rodent from East Africa " by D. G. MacInnes. 1957, £1.

No. 13. " Insectivora and Chiroptera from the Miocene Rocks of Kenya Colony " by P. M. Butler and A. Tindell Hopwood. 1957, 15s.

No. 14. " Some East African Pleistocene Suidæ " by L. S. B. Leakey. 1958, £3 10s

No. 15. " Miocene Ruminants of East Africa " by T. Whitworth. 1958, 14s.

No. 16. "The Fore-Limb Skeleton and Associated Remains of Proconsul africanus" by J. R. Napier and P. R. Davis. 1959, £1 15s.

No. 17. "Fossil Metacarpus from Swartkrams" by J. R. Napier. 1959, 8s.

No. 18. "East Africa Miocene and Pleistocene Chalicotheres" by P. M. Butler. 1965, £1 8s.

No. 19. "The Miocene Carnivora of East Africa" by R. J. G. Savage. 1965, £2 12s.

No. 20. "Fossil Antilopini of East Africa" by A. W. Gentry. 1966, £2 5s.

No. 21. "Miocene Rhinoceroses of East Africa" by D. A. Hooijer. 1966, £3 3s.

No. 22. "Pelorovis oldowayensis Reck, an Extinct Bovid from East Africa" by A. W. Gentry. 1967, £1 16s.

Published by the British Museum
(Natural History), London.